Springer Tracts in Modern Physics
Volume 135

Managing Editor: G. Höhler, Karlsruhe

Editors: J. Kühn, Karlsruhe
Th. Müller, Karlsruhe
R. D. Peccei, Los Angeles
F. Steiner, Ulm
J. Trümper, Garching
P. Wölfle, Karlsruhe

Honorary Editor: E. A. Niekisch, Jülich

Springer
Berlin
Heidelberg
New York
Barcelona
Budapest
Hong Kong
London
Milan
Paris
Santa Clara
Singapore
Tokyo

Springer Tracts in Modern Physics

Volumes 120–136 are listed at the end of the book

Covering reviews with emphasis on the fields of Elementary Particle Physics, Solid-State Physics, Complex Systems, and Fundamental Astrophysics

Manuscripts for publication should be addressed to the editor mainly responsible for the field concerned:

Gerhard Höhler
Institut für Theoretische Teilchenphysik
Universität Karlsruhe
Postfach 6980
D-76128 Karlsruhe
Germany
Fax: +49 (7 21) 37 07 26
Phone: +49 (7 21) 6 08 33 75
Email: hoehler@fphvax.physik.uni-karlsruhe.de

Johann Kühn
Institut für Theoretische Teilchenphysik
Universität Karlsruhe
Postfach 6980
D-76128 Karlsruhe
Germany
Fax: +49 (7 21) 37 07 26
Phone: +49 (7 21) 6 08 33 72
Email: johann.kuehn@physik.uni-karlsruhe.de

Thomas Müller
IEKP
Fakultät für Physik
Universität Karlsruhe
Postfach 6980
D-76128 Karlsruhe
Germany
Fax:+49 (7 21) 6 07 26 21
Phone: +49 (7 21) 6 08 35 24
Email: mullerth@vxcern.cern.ch

Roberto Peccei
Department of Physics
University of California, Los Angeles
405 Hilgard Avenue
Los Angeles, California 90024-1547
USA
Fax: +1 310 825 9368
Phone: +1 310 825 1042
Email: robertop@college.ucla.edu

Frank Steiner
Abteilung für Theoretische Physik
Universität Ulm
Albert-Einstein-Allee 11
D-89069 Ulm
Germany
Fax: +49 (7 31) 5 02 29 24
Phone: +49 (7 31) 5 02 29 10
Email: steiner@physik.uni-ulm.de

Joachim Trümper
Max-Planck-Institut
für Extraterrestrische Physik
Postfach 1603
D-85740 Garching
Germany
Fax: +49 (89) 32 99 35 69
Phone: +49 (89) 32 99 35 59
Email: jtrumper@mpe-garching.mpg.de

Peter Wölfle
Institut für Theorie
der Kondensierten Materie
Universität Karlsruhe
Postfach 69 80
D-76128 Karlsruhe
Germany
Fax: +49 (7 21) 69 81 50
Phone: +49 (7 21) 6 08 35 90/33 67
Email: woelfle@tkm.physik.uni-karlsruhe.de

Alois Würger

From Coherent Tunneling to Relaxation

Dissipative Quantum Dynamics
of Interacting Defects

With 51 Figures

 Springer

Dr. Alois Würger
Institut Max von Laue-Paul Langevin
Avenue des Martyrs, B.P. 156
F-38042 Grenoble Cedex 9
France
Email: wuerger@ill.fr

Cataloging-in-Publication Data applied for

Die Deutsche Bibliothek - CIP-Einheitsaufnahme

Würger, Alois:
From coherent tunneling to relaxation : dissipative quantum dynamics of interacting defects / Alois Würger. - Berlin ; Heidelberg ; New York ; Barcelona ; Budapest ; Hong Kong ; London ; Milan ; Paris ; Santa Clara ; Singapore ; Tokyo : Springer, 1997
(Springer tracts in modern physics ; Vol 135)
ISBN 3-540-61424-9
NE: GT

Physics and Astronomy Classification Scheme (PACS): 62, 65, 66

ISSN 0081-3869
ISBN 3-540-61424-9 Springer-Verlag Berlin Heidelberg New York

This work is subject to copyright. All rights are reserved, whether the whole or part of the material is concerned, specifically the rights of translation, reprinting, reuse of illustrations, recitation, broadcasting, reproduction on microfilm or in any other way, and storage in data banks. Duplication of this publication or parts thereof is permitted only under the provisions of the German Copyright Law of September 9, 1965, in its current version, and permission for use must always be obtained from Springer-Verlag. Violations are liable for prosecution under the German Copyright Law.

© Springer-Verlag Berlin Heidelberg 1997
Printed in Germany

The use of general descriptive names, registered names, trademarks, etc. in this publication does not imply, even in the absence of a specific statement, that such names are exempt from the relevant protective laws and regulations and therefore free for general use.

Typesetting: Camera-ready copy from the authors using a Springer T$_E$X macro package.
Cover design: Springer-Verlag, Design & Production
SPIN: 10528107   56/3144-5 4 3 2 1 0 – Printed on acid-free paper

Foreword

Tunneling of electrons through potential barriers is a typical quantum mechanical phenomenon and has fundamental consequences for the properties of solids. In contrast, tunneling of entire atoms or molecules seems to be of minor importance. Surprisingly, however, the low-temperature properties of real solids can be determined to a large extent by just this phenomenon. A well-known example for atomic tunneling is the motion of nitrogen atoms in ammonium molecules. This process was discussed as early as 1927 by F. Hund shortly after the foundation of quantum mechanics and is the basis of the ammonium maser realized much later. The tunneling of atoms in solids was first addressed by L. Pauling in 1930, but was not clearly verified experimentally before 1962 when W. Känzig was able to show that at low temperatures O_2^- ions perform tunneling motion in alkalihalides. Since this milestone many theoretical and experimental studies have been carried out on similar systems and it now seems that the fundamental questions about atomic tunneling in solids could be answered within a short time.

A far-reaching change occurred in this field of research when the pioneering work of R.C. Zeller and R.O. Pohl was published in 1971. These authors demonstrated that glasses exhibit anomalous thermal properties at low temperatures. Shortly afterwards a phenomenological description of the surprising observations was developed independently by W.A. Phillips and by P.W. Anderson, B.I. Halperin and C. Varma. This so-called 'tunneling model' is based on the assumption that "particles" exist in amorphous structures, which are able to tunnel between different potential minima. The new field immediately attracted much interest and less attention was paid to the properties of tunneling states in crystals. However, it turned out that the progress made in understanding tunneling processes in amorphous solids has been much slower than it was in crystals, mainly due to the difficulties inherently connected with the theoretical treatment of random structures. Although the tunneling model is an important guide for further theoretical developments and allows a good phenomenological description of many low-temperature properties of glasses, the universality of the observed phenomena and the microscopic origin of the tunneling systems is still not well understood.

There is no doubt that a close relationship exists between the phenomena observed in crystals and amorphous solids. Therefore efforts were made to find experimental and theoretical links. One such promising idea followed by several groups was the investigation of crystalline "model substances", i.e., of crystals containing a higher concentration of tunneling systems. A well-known example is the mixed crystal KBr:KCN, in which tunneling systems are formed by CN^- ions. At low concentrations the tunneling systems can be considered to be isolated. With increasing concentration the "glassy behavior" of the crystal becomes more and more pronounced. For example, if 25% of the bromium ions are replaced by CN^- ions, the specific heat increases linearly with temperature like in glasses and reaches a magnitude very close to that of vitreous silica. At a first glance this approach seems to be very promising. However, the "glassy behavior" is not caused by isolated tunneling states but is due to the interaction between all the systems, i.e., we face a complex many-body problem. A solution to this problem is still missing even for the model substances, and only a phenomenological description of experimental results can be achieved so far, as in the case of glasses. In this book, however, the interaction between tunneling systems has been taken into account in a theory going beyond mean field approximations. Although the theory given here has been worked out for tunneling systems interacting via electrical dipole forces, there is hope that it can be extended to the more complicated elastic interaction, which is the dominant interaction in the model substances.

This book reflects the important and successful attempt of the author to develop new theoretical concepts that can be applied to the tunneling systems of crystals and amorphous solids as well. In particular, the interaction of the tunneling systems with the environment is taken into account and plays a central role. The comprehensive treatment of tunneling by atoms and small clusters is contained in two chapters that introduce the reader to those theories currently used. In one chapter the influence of isolated defects on low-temperature properties of crystals is discussed; in the other (at the beginning of the description of anomalous properties of amorphous solids), the well-known tunneling model is introduced. Later on, the two concepts serve as starting points for the description of novel and exciting theoretical developments that have very recently been proposed by the author and his coworkers. An aspect of fundamental importance is the treatment of the cross-over from coherent to incoherent tunneling. This transition occurs if the interaction of the tunneling systems with their environment becomes sufficiently strong. For crystals it is the interaction between the tunneling systems that destroys the phase of the tunneling particle and opens a new channel of relaxation. Obviously the importance of this effect increases with increasing concentration of interacting systems and gives rise to many surprising observations. It is the first time that a consistent theory has been developed for their description.

In amorphous solids the situation is more complex. From the observation of phenomena that are caused by the coherent motion of tunneling particles one can conclude that interaction between tunneling systems is of minor importance. Nevertheless a cross-over from coherent tunneling to relaxation occurs in glasses as well. As shown by the author in this case the phase of the tunneling particle is destroyed by the thermal motion of the neighbors, i.e., by the interaction of tunneling states with the phonon bath.

As an experimentally oriented physicist I especially appreciate that not just interesting theoretical concepts and developments are presented in this book. The author always keeps in mind that his theoretical considerations are related to experimental observations and in many cases the experimental facts are even his starting point. Without doubt this book is not only a pleasure for theorists but also for experimentalists.

I want to take the opportunity to thank Dr. Alois Würger for the close collaboration with my group and for the many hours he spent having discussions in our laboratory during the last years. I think it was a very stimulating and fruitful time for both sides.

Heidelberg, June 1996   Siegfried Hunklinger

Preface

Most of the work presented in this book grew out of the cooperation of groups at the Institut für Theoretische Physik and the Institut für Angewandte Physik of the Ruprecht-Karls-Universität Heidelberg. I am most grateful to Dr. Christian Enss for our pleasant collaboration during the last years. In particular, I thank him and Dr. Robert Weis for their kind permission to reproduce here various – partly unpublished – experimental data.

At the Institut für Theoretische Physik, Dr. Orestis Terzidis and Dr. Peter Neu have contributed greatly to the results reported in Chaps. 4 and 7 of this book, and I am indebted to them for many critical and stimulating discussions.

Finally, I would like to thank Prof. S. Hunklinger and Prof. H. Horner for their support and encouragement.

Grenoble, August 1996   Alois Würger

Contents

1. **Introduction** 1
   1.1 Substitutional Defects in Alkali Halide Crystals 3
   1.2 Tunneling Systems in Amorphous Solids 6

2. **Tunneling of a Single Defect** 9
   2.1 Energy Spectrum and Tunneling States 9
   2.2 Dielectric Response Function 14
   2.3 Elastic Response Function 16
   2.4 Thermal Conductivity 25
   2.5 Summary 26

3. **Tunneling of a Coupled Defect Pair** 27
   3.1 Perturbation Theory 27
   3.2 Configuration NN2 31
   3.3 Configuration NN1 37
   3.4 Configuration NN3 39
   3.5 Rotary Echoes 40
   3.6 Discussion 46

4. **A Pair of Two-Level Systems** 51
   4.1 Two-State Approximation 51
   4.2 The Pair Model 53
   4.3 The Statistical Operator 54
   4.4 Density of States 57
   4.5 Specific Heat 58
   4.6 Dynamic Susceptibility 61

5. **Cross-Over to Relaxation: Mori Theory** 67
   5.1 Continued Fraction Expansion 69
   5.2 Mode-Coupling Approximation 77
   5.3 Markov Approximation 80
   5.4 Mean-Field Approximation for κ^2 82
   5.5 Density of States 84
   5.6 Specific Heat 87

	5.7 Resonant Susceptibility 92
	5.8 Relaxational Susceptibility 96
	5.9 Discussion 107
	5.10 Summary and Outlook 112

6. The Tunneling Model 115
   6.1 Low-Temperature Properties of Glasses 115
   6.2 The Tunneling Model 118
   6.3 Distribution of the Parameters Δ_0 and Δ 121
   6.4 The Heat Bath 122

7. Tunneling Systems in Amorphous Solids 125
   7.1 Coherent Tunneling: Perturbation Theory 126
   7.2 Incoherent Tunneling: Mode-Coupling Theory 129
   7.3 Dynamics Beyond the Two-Level Approximation 135
   7.4 Sound Propagation 136
   7.5 Comparison with Experimental Results.................... 142
   7.6 Summary... 149

8. Small-Polaron Approach 153
   8.1 Break-Down of Perturbation Theory 153
   8.2 Small-Polaron Representation 154
   8.3 Limit of Zero Damping 157
   8.4 Approximation Scheme................................. 158
   8.5 The Damping Rate Γ_1 163
   8.6 Cross-Over to Incoherent Motion 168
   8.7 The Correction Term Σ_2 170
   8.8 Discussion 177
   8.9 Summary and Conclusion 183

A. Spectral Representation 185

Appendices 185

B. Mori's Reduction Method 187
   B.1 Scalar Product 187
   B.2 Correlation Matrix of Relevant Operators 188
   B.3 Continued Fraction Expansion 189
   B.4 Approximations 190

C. Average over Disorder 193
   C.1 Distribution of Dipolar Couplings J_{ij} 193
   C.2 The Distribution Function $Q(I)$ 195
   C.3 Distribution of τ_i 197

D. The Zero-Frequency Limit of $K_i^I(z)$. 199

E. Small-Polaron Approach 203
  E.1 Time Correlation Functions of Polaron Operators 203
  E.2 Upper and Lower Bounds for Γ_1 204

References 207

Index 213

1. Introduction

Tunneling between degenerate states has been a controversial topic since the early days of quantum mechanics. In 1927, Hund investigated as an illustrative example the resonance spectrum of the ammonia molecule NH_3 [1], which may be found in two equivalent positions for the nitrogen, to the left and to the right of the plane formed by the hydrogen atoms. This inversion symmetry gives rise to a ground state doublet with a tiny tunnel splitting $\hbar\omega_t$; the energy eigenstates are even and odd superpositions of the two molecular configurations. When it starts from one of the localized states, and if there are no perturbations, the ammonia molecule performs coherent oscillations between the two positions with the tunnel frequency $\omega_t/2\pi$. Subsequent microwave absorption experiments provided a value of $\omega_t/2\pi = 2.4 \times 10^{10}$ Hz for the oscillation frequency [2,3].

In biology certain optically active molecules are found to occur in one chiral configuration only; Nature does not create the mirror molecule. The persistence of optical activity over long periods of time shows that there are no tunneling transitions between the two configurations. Hund explained this 'paradox of optical isomeres' by the exponential dependence of the tunnel frequency on the number of atoms involved in the configurational reorientation; in a rough numerical estimate he obtained oscillation periods of up to 10^9 years [1]. Thus tunneling is unlikely for large and heavy systems. More recently the advent of small-scale superconducting devices revived the interest in and stimulated the search for quantum coherence on a macroscopic scale, related e. g. to the Josephson effect or flux quantization in a superconducting ring [4]. Yet it seems that so far no coherent superposition of macroscopic states has been observed.

There are two sources for this lack of coherence, namely the exponentially small tunnel frequency, and the perturbation of time evolution by the thermal motion of the environment. Yet even for small molecules with a sufficiently large tunnel frequency, the thermal motion of the surrounding atoms may destroy the phase coherence of tunneling oscillations. The inversion resonance of gaseous ammonia has been measured at room temperature [3]. Quantum coherence in solids requires much lower temperatures; as for tunneling systems arising from molecular defects in crystals, no phase coherence has been found above about 50 K. From this we may conclude that quantum coherence is most likely to be observed for light systems at very low temperatures, thus ruling out situations such as that shown below.

Quantum coherence or *coherent tunneling* arises from particles moving back and forth between degenerate states, such as Bloch electrons in a periodic potential, or inversion resonance of an ammonia molecule. This effect is to be distinguished from the escape of a particle out of a metastable state by tunneling through a potential barrier; as examples we note nuclear fission with emission of an alpha particle and the phase motion of a superconducting ring interrupted by a Josephson junction, the latter involving a truly macroscopic coordinate. Since after it escapes, the system never returns to its original state, it does not interfere with itself [4]. Superconducting devices have become a standard probe for macroscopic quantum effects; the superconducting phase provides the macroscopic coordinate, and the tunnel effect causes a phase slip of 2π [5].

This work addresses quantum coherence on a microscopic scale emerging from the tunneling motion of a single atom or molecule, or a small group of atoms, between degenerate quantum states; dissipation arises from interaction with either an oscillator heat bath or other tunneling systems. The simplest model consists of an atom which has two positions in a double-well potential; the tunnel effect causes a leaking of the probability amplitude from one well into the other, and the heat bath induces fluctuations of the shape and the relative depth of the two wells.

The question about quantum coherence amounts to asking whether a particle that started in the left well will move to and fro periodically in time, or whether it will have lost its phase memory when reaching the right-hand well. In the first case we have coherent tunneling with a well-defined tunneling frequency, and the particle's motion is best described in terms of even and odd energy eigenstates. In the second one, the motion is incoherent in nature and characterized by a relaxation rate defined as the inverse barrier crossing time. The basic problem consists in properly describing the cross-over between the two regimes and in calculating the relaxation rate from some model Hamiltonian.

A two-level system coupled to a bosonic heat bath with a power-law spectral density has become popular as the spin-boson model, and it has been applied to various situations in physics and chemistry [6,7]. During the last decade, a lot of work was devoted to the case of a linear spectrum, or Ohmic dissipation, and to its many applications [6–25]. (At finite temperatures a linear spectral density leads to a frequency-independent 'Ohmic' damping rate.) In the second part of this book (Chaps. 6–8) we are concerned with the spin-boson model with a cubic spectral density as it arises from coupling to lattice vibrations; the first part (Chaps. 2–5) deals with interacting tunneling systems. In the remainder of this introductory chapter, we briefly sketch the contents.

1.1 Substitutional Defects in Alkali Halide Crystals

Certain impurity ions in alkali halides form para-electric off-center tunneling systems which are characterized by a ground-state splitting Δ_0 and a dipole moment p [29–32]. In Fig. 1.1 we show a lithium impurity in potassium chloride (KCl:Li); as further examples we note NaCl:OH, and KBr:CN. The tunnel energy takes values cprresponding to about 1 K. For N impurities in a sample of volume V we find the defect density $n \equiv N/V$, and the mean distance of adjacent impurities is $n^{-1/3}$. When we discard a numerical factor and the dielectric constant, the average dipolar interaction is given by np^2.

Thus there are two energy scales for interacting impurities, and the thermal and dynamic behavior turns out to be different for the cases $\Delta_0 \ll np^2$ and $np^2 \ll \Delta_0$. The cross-over condition $\Delta_0 = np^2$ obviously depends on tunnel energy and dipole moment; inserting typical values one finds it to be satisfied by an impurity concentration of about 1000 ppm, corresponding to a density n of about 10^{19} cm^{-3}.

In the dilute limit the mean distance $n^{-1/3}$ is large, and the average interaction energy np^2 is much smaller than the tunnel splitting. This situation is shown schematically in the upper part of Fig. 1.2. Impurities are represented by dots, and we have defined an interaction radius R by taking $p^2/R^3 = \Delta_0$;

thus for two defects at distance R, the dipolar interaction is equal to the tunnel energy. Clearly at very low concentrations the relation $R \ll n^{-1/3}$ renders the interaction insignificant, and the impurities may be considered as independent from each other. The single-impurity case is thoroughly discussed in Chap. 2, where we derive the energy spectrum and selection rules for electric dipole and elastic quadrupole transitions, and calculate the corresponding dynamical response functions.

When increasing the density n, most impurities may still be considered as non-interacting; yet due to the random distribution on the host lattice a few of them get sufficiently close and form coupled pairs, as indicated in the middle part of Fig. 1.2 by overlapping interaction circles. There are two relevant aspects arising from interaction.

First, although occurring with very small probability, the particular dynamics of two impurities on nearest-neighbor or next-nearest-neighbor sites may be detected by an appropriate experimental means. In Chap. 3 we investigate such special pair configurations and discuss closely related rotary echo experiments.

Second, calculation of dielectric susceptibility and specific heat requires an average over all configurations. At sufficiently low densities, i. e., $R \ll n^{-1/3}$, a virial expansion provides a description in terms of powers of n. The first non-trivial order yields pairs of coupled impurities with a continuous distribution for the interaction energy; the N-particle problem thus factorizes in $N/2$ pairs. In a second approximation, the actual eight-state system describing a single defect is replaced by a two-state system. One thus obtains pairs of two-level systems whose static and dynamic properties are compared with observed data in Chap. 4.

When we further increase the impurity density, the break-down of the pair model occurs at $n^{-1/3} \approx R$, or $np^2 \approx \Delta_0$. The virial expansion is not a good starting point, since clusters consisting of three or more defects are not

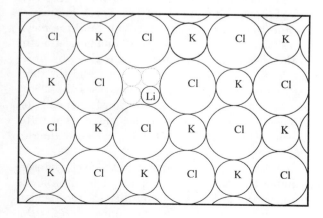

Fig. 1.1. Off-center positions of a lithium impurity, projected on the $x - y$ plane. There are four impurity sites above the plane shown, and four sites below

1.1 Substitutional Defects in Alkali Halide Crystals

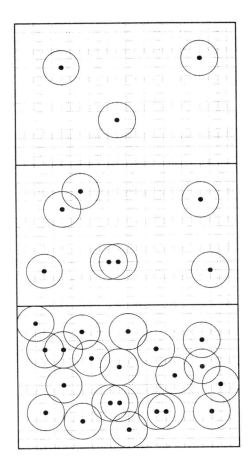

Fig. 1.2. Random configuration of N lithium impurities on the host lattice. Each black point represents one impurity as in Fig. 1.1. The circles indicate the 'interaction radius' R; for two impurities whose circles touch, the interaction p^2/R^3 is equal to the tunnel energy Δ_0

negligible, and the state of N impurities no longer factorizes in $N/2$ pairs, as is illustrated in the lower part of Fig. 1.2. This intricate many-body problem is treated in Chap. 5 by means of Mori's reduction method, which permits us to derive a continued fraction representation for the relevant two-times correlation function.

We construct two different approximation schemes that lead to correlation spectra comprising three and five poles, respectively. In the dilute limit, both schemes tend towards the low-density results of Chap. 4, which are characterized by coherent tunneling of one or two impurities. For the opposite case of high concentrations (i. e., above about 1000 ppm), this approach yields purely relaxational dynamics with a broad distribution of rates. Applying a mode-coupling approximation, we obtain equations that resemble those derived for spin glass and glass dynamics.

The main results of Chap. 5 are the emergence of a strong relaxation peak and a *decrease* in the dielectric susceptibility when the defect density is augmented beyond $np^2 \approx \Delta_0$. Both features agree with experimental findings.

1.2 Tunneling Systems in Amorphous Solids

Quantum tunneling of atoms is a ubiquitous phenomenon in disordered solids. Examples are substitutional defects as discussed above, interstitial hydrogen in metals, and materials with configurational disorder such as insulating glasses, polymers, and amorphous metals.

For substitutional and interstitial defects, specific microscopic models have been developed whose parameters may be induced from experiment; in a few cases these models could be justified by ab initio calculations with realistic atom-atom potentials.

The situation is very different for tunneling systems in amorphous solids. The lack of long-range order in general prohibits a proper characterization of the atomic potential energy landscape. Moreover, the tunneling systems do not correspond to well-defined atoms or molecules, but rather involve the – more or less random – local configuration.

On the other hand, the similar properties of various amorphous materials at low temperatures seem to derive from the same physical cause. The stiff network of directed covalent bondings in oxide glasses, the rather one-dimensional structure of polymers, and the non-stoichiometric alloys forming metallic glasses have little in common on an atomic level. Thus searching for a microscopic explanation for universal behavior at low temperatures would seem a pointless undertaking. Up to now the most successful attempt is given by the tunneling model, which we discuss in Chap. 6.

Various experiments provide conclusive evidence for the existence of two-level systems with a wide range of energy splittings whose time evolution may be strongly affected by coupling to low-frequency lattice vibrations. The latter cause potential fluctuations; in terms of a double-well potential they induce a time-dependent asymmetry energy which tends to destroy the phase coherence of two-state oscillations. Accordingly, there are two relevant frequencies, the first one, $E/\hbar$, being determined by the two-level splitting E, and the second one by the phonon damping rate Γ.

At sufficiently low temperatures, most phonon modes are frozen out, and the two-level system performs weakly damped oscillations with frequency $E/\hbar$. Increasing the temperature enhances the thermal lattice motion which, in turn, leads to fluctuations of the relative depth of the two wells. At some point these potential fluctuations exceed the two-level splitting E and thus wipe out coherent time evolution.

Clearly the basic problem is to calculate the effective damping rate as a function of temperature and to specify the dynamic behavior in the incoherent regime. In Chap. 7 we present a mode-coupling approach to the spin-boson model. The solution of the resulting set of integral equations for the longitudinal and transverse correlation functions of a two-level system exhibits a cross-over from coherent tunneling to relaxation with rising temperatures. Finally, we compare the results for sound velocity and attenuation with experimental data on various amorphous materials.

The mode-coupling approach neglects the phonon dressing of the tunneling particle, which results in an effective mass or a renormalized tunnel energy. In Chap. 8 we treat the spin-boson model using a different technique, which fully accounts for the lattice distortion accompanying the particle dwelling in the left or right well; its essential approximation corresponds roughly to a power series expansion with respect to the tunnel energy.

2. Tunneling of a Single Defect

2.1 Energy Spectrum and Tunneling States

Substitutional defects in alkali halides provide a model system for the study of quantum tunneling in a crystalline environment. As examples we note Li, CN, and OH in potassium chloride and similar crystals. Tunneling arises from defects which do not fit properly in the sites offered by the host lattice.

The potential experienced by the defect ion is mainly determined by the symmetry of the host crystal. Potassium chloride shows a fcc structure with two positive ions per unit cell; nearest-neighbor potassium sites are at a distance of 4.5 Å. The ionic radius of a lithium defect, $r_{Li} = 0.60$ Å, is by a factor of two smaller than that of the potassium ion for which it substitutes, $r_K = 1.33$ Å; thus the cage formed by the six nearest chloride ions is too large for a lithium ion. Whereas the lattice vertex constitutes the energetically most favorable position for K^+, the smaller lithium ion encounters a more

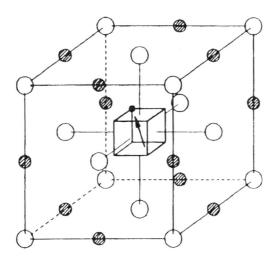

Fig. 2.1. Off-center positions of a Lithium impurity in KCl. Ion radii are reduced by a factor of $\frac{1}{5}$ for visibility. Open, hatched, and full circles represent Cl, K, and Li ions, respectively. The small cube indicates the eight Li positions; the arrow shows the direction of the dipole moment. By courtesy of M. Tornow [99]

complicated potential energy landscape, which in particular exhibits several degenerate minima at so-called off-center positions.

Three different sets of defect positions are compatible with the host's symmetry. The first involves eight minima in [111]-directions; the second and the third exhibit six minima on the [100] axes and twelve minima in [110]-directions, respectively. Quantum tunneling leads to a characteristic ground-state splitting for each of these situations; the resulting energy spectra and tunneling states have been discussed in detail by Gomez et al. [32]. Comparison with measured data permitted us to conclude that in KCl:Li the first of the mentioned possibilities is achieved; the resulting eight off-center positions in [111] directions form a cube of side-length $d = 1.4$ Å about the original lattice site. (Compare Fig. 2.1.)

The simplest model consists of quadratic potentials which match at the planes $x = 0$, $y = 0$, and $z = 0$,

$$V(x,y,z) = \frac{1}{2}m\omega^2 \left[(|x| - d/2)^2 + (|y| - d/2)^2 + (|z| - d/2)^2\right], \quad (2.1)$$

with minima symmetrically located on the axes at a distance of $\frac{1}{2}d$ from the origin. Tunneling occurs along the edges of the cube and along face and space diagonals; for the distance between corresponding minima we find values of d, $\sqrt{2}d$, and $\sqrt{3}d$, respectively. The states localized at

$$\boldsymbol{r} = \tfrac{1}{2}d(\alpha, \beta, \gamma) \quad (2.2)$$

are denoted by $|\alpha\beta\gamma\rangle$. (Greek indices take the values ± 1.)

Accordingly, there are three different tunneling amplitudes

$$\begin{aligned}
-\tfrac{1}{2}\Delta_\mathrm{K} &= \langle +++|V|++-\rangle, \\
-\tfrac{1}{2}\Delta_\mathrm{F} &= \langle +++|V|+--\rangle, \\
-\tfrac{1}{2}\Delta_\mathrm{R} &= \langle +++|V|---\rangle;
\end{aligned} \quad (2.3)$$

these are identical for all pairs of states differing by one, two, or three signs. (The factor $\tfrac{1}{2}$ has become usual; in order to form a link with the notation of [32], put $\Delta_\mathrm{K} \to 2\eta$, $\Delta_\mathrm{F} \to 2\mu$, and $\Delta_\mathrm{R} \to 2\nu$.) The localized states $|\alpha\beta\gamma\rangle$ are

Table 2.1. Tunnel energies for ^{7}Li, ^{6}Li, and CN impurities in KCl. Lorentz correction of the bare dipole moment p_0 yields the effective value p. Cited after [41]

	Δ_K / K	Δ_F / K	Δ_R / K	p_0 / D	p / D
^{7}Li [68]	1.10	-0.34	0.20	5.68	2.63
^{6}Li [69]	1.65			5.60	2.59
CN [70,71]	1.60				0.30

not perpendicular to each other; yet with $\langle\alpha\beta\gamma|\alpha\beta\gamma\rangle = 1$ and for sufficiently deep potential wells, the off-diagonal overlap matrix elements

$$\begin{aligned} S_{\rm K} &= \langle +++|++-\rangle, \\ S_{\rm F} &= \langle +++|+--\rangle, \\ S_{\rm R} &= \langle +++|---\rangle, \end{aligned} \quad (2.4)$$

are much smaller than unity. The tunneling spectrum is entirely determined by the six parameters (2.3,2.4), which may be calculated by means of an appropriate ansatz for the localized wave functions; in many instances, modified harmonic oscillator functions

$$\langle \alpha\beta\gamma|xyz\rangle = \left(\frac{m\omega}{\pi\hbar}\right)^{3/4} \exp\left[-\frac{m\omega}{2\hbar}(\hat{\boldsymbol{r}} - \boldsymbol{r})^2\right], \quad (2.5)$$

with (2.2) and the position operator $\hat{\boldsymbol{r}} = (x, y, z)$, turned out to provide a proper description. (Note that in the sequel, the two-state coordinates (2.2) are referred to as position.)

For a sufficiently high barrier, the spread of the wave function (2.5) is much smaller than the distance between the two minima; because of

$$m\omega d^2/\hbar \gg 1,$$

the tunneling amplitudes may then be evaluated in semiclassical approximation [81]. We note the simple WKB result $\hbar\omega_0 \exp(-d\sqrt{2mV}/\hbar)$ for tunneling through a barrier of height V and width d; from its distance-dependence, one is led to the relation

$$\Delta_{\rm R} \ll \Delta_{\rm F} \ll \Delta_{\rm K}. \quad (2.6)$$

In the limit of deep potential wells, the overlap integrals (2.4) of different localized states $|\alpha\beta\gamma\rangle$ fulfil

$$S_{\rm R} \ll S_{\rm F} \ll S_{\rm K} \ll 1, \quad (2.7)$$

and thus will be neglected in the sequel; moreover the continuous coordinates $\hat{\boldsymbol{r}}$ can be replaced by the three effective two-state coordinates $\boldsymbol{r}$ as given in (2.2).

Here we will not attempt such a calculation from a microscopic potential, but rather consider the tunneling amplitudes (2.3) as parameters that can be deduced from experiment. In Table 2.1 we show values for the tunnel amplitudes and the dipole moment, which have been determined by paraelectric resonance and specific heat measurements. As to the largest energy $\Delta_{\rm K}$, values reported by various authors confirm those given in Table 2.1. Yet for the smaller energies $\Delta_{\rm F}$ and $\Delta_{\rm R}$, only a few data are available, and these are subject to great uncertainty and are often contradictory. (See e.g. [69], where much smaller values for $\Delta_{\rm F}$ and $\Delta_{\rm R}$ are obtained.)

According to (2.3), the tunneling Hamiltonian consists of three parts, corresponding to tunneling along the edges and along the face and space diagonals of the cube,

2. Tunneling of a Single Defect

$$H = T_\text{K} + T_\text{F} + T_\text{R}; \tag{2.8}$$

using standard basis operators they read

$$
\begin{aligned}
T_\text{K} &= -\frac{\Delta_\text{K}}{2} \sum_{\alpha,\beta,\gamma} \Big(|\alpha\beta\gamma\rangle\langle\alpha\beta\bar\gamma| + |\alpha\beta\gamma\rangle\langle\alpha\bar\beta\gamma| + |\alpha\beta\gamma\rangle\langle\bar\alpha\beta\gamma| \Big), \\
T_\text{F} &= -\frac{\Delta_\text{F}}{2} \sum_{\alpha,\beta,\gamma} \Big(|\alpha\beta\gamma\rangle\langle\alpha\bar\beta\bar\gamma| + |\alpha\beta\gamma\rangle\langle\bar\alpha\bar\beta\gamma| + |\alpha\beta\gamma\rangle\langle\bar\alpha\beta\bar\gamma| \Big), \\
T_\text{R} &= -\frac{\Delta_\text{R}}{2} \sum_{\alpha,\beta,\gamma} |\alpha\beta\gamma\rangle\langle\bar\alpha\bar\beta\bar\gamma|,
\end{aligned}
\tag{2.9}
$$

where α, β, γ take the values ± 1 and $\bar\alpha = -\alpha$. The energy spectrum has been investigated in detail in [32]; here we discuss only the simple case where Δ_F and Δ_R are negligible as compared to Δ_K.

The energy eigenstates are irreducible representations of the point group O_h; they are constructed as superpositions of the localized states,

$$|\boldsymbol{\sigma}\rangle = 8^{-1/2} \sum_{\{\alpha_i = \pm 1\}} f_{\alpha_x}^{\sigma_x} f_{\alpha_y}^{\sigma_y} f_{\alpha_z}^{\sigma_z} |\alpha_x \alpha_y \alpha_z\rangle, \tag{2.10}$$

with appropriate phase factors

$$f_\alpha^\sigma = \exp\left(\frac{\text{i}\pi}{4} (\sigma+1)(\alpha+1) \right). \tag{2.11}$$

The entries of the vector $\boldsymbol{\sigma} = (\sigma_x, \sigma_y, \sigma_z)$ take the values ± 1; thus we have $f_\alpha^\sigma = -1$ for $\alpha = 1 = \sigma$ and $f_\alpha^\sigma = 1$ else.

The eight states (2.10) are arranged in four levels; indicating both the corresponding representation of O_h and the label $\boldsymbol{\sigma}$, and using $\bar 1 = -1$, we have

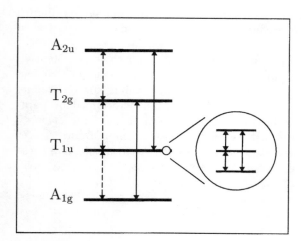

Fig. 2.2. Energy levels of a lithium ion in KCl according to (2.12). The dashed arrows indicate dipole transitions, and the solid arrows are those mediated by elastic strain. Compare (2.20) and (2.35)

$$\begin{aligned}
E(\mathrm{A_{2u}}) &= \tfrac{3}{2}\Delta_\mathrm{K} &&: (111), \\
E(\mathrm{T_{2g}}) &= \tfrac{1}{2}\Delta_\mathrm{K} &&: (11\bar{1}),(1\bar{1}1),(\bar{1}11), \\
E(\mathrm{T_{1u}}) &= -\tfrac{1}{2}\Delta_\mathrm{K} &&: (1\bar{1}\bar{1}),(\bar{1}1\bar{1}),(\bar{1}\bar{1}1), \\
E(\mathrm{A_{1g}}) &= -\tfrac{3}{2}\Delta_\mathrm{K} &&: (\bar{1}\bar{1}\bar{1}).
\end{aligned} \qquad (2.12)$$

The labels g and u indicate even and odd parity; the levels with T-symmetry are threefold degenerate.

Dynamic quantities depend crucially on the selection rules for transitions between the energy eigenstates. For a spatially slowly varying electric field dipole transitions are predominant, whereas an elastic strain field couples mainly to the quadrupole moment. In Fig. 2.2 we plot the spectrum with both labels and we indicate the allowed dipole and quadrupole transitions.

For $\Delta_\mathrm{F} = 0 = \Delta_\mathrm{R}$, both partition function $Z = \mathrm{tr}(e^{-\beta H})$ and internal energy $U = \langle H \rangle = -\partial_\beta \log(Z)$ take a particularly simple form. The former is given by the third power of a two-level partition function

$$Z = 2^3 \cosh(\beta \Delta_\mathrm{K}/2)^3; \qquad (2.13)$$

defining the thermal average

$$\langle ... \rangle = Z^{-1} \mathrm{tr}(...e^{-\beta H}) \qquad (2.14)$$

and $k_\mathrm{B} T = 1/\beta$, the internal energy for a single impurity ion reads

$$U = -\tfrac{3}{2}\Delta_\mathrm{K} \tanh(\Delta_\mathrm{K}/2k_\mathrm{B} T). \qquad (2.15)$$

When we consider N defects in a sample of volume V, we find with the number density $n = N/V$ the specific heat

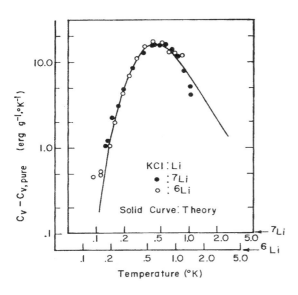

Fig. 2.3. Specific heat of lithium impurities in KCl. The Debye contribution of the host crystal has been subtracted. Note the different temperature scale for the two isotopes ^{6}Li and ^{7}Li. The solid line is given by (2.16) with $\Delta_\mathrm{K}/k_\mathrm{B} = 1.16$ K for ^{7}Li and $\Delta_\mathrm{K}/k_\mathrm{B} = 1.62$ K for ^{6}Li; the defect density used for the fits deviates from that determined by spectro-chemical analysis. From [57]

$$C_V = 3nk_B(\Delta_K/2k_BT)^2 \cosh(\Delta_K/2k_BT)^{-2}, \tag{2.16}$$

which exhibits a characteristic Schottky anomaly at a temperature $T \approx \Delta_K/2k_B$. Fig. 2.3 reveals a perfect agreement of (2.16) with respect to the isotope effect. (The values for the tunnel energies $^7\Delta_K/k_B = 1.16$ K and $^6\Delta_K/k_B = 1.64$ K obtained by Harrison et al. [57] slightly differ from those used in the sequel of this book (compare Table 2.1). Yet the defect density obtained from the fits in Fig. 2.3 does not quite agree with the value determined by spectro-chemical analysis; possible sources for this inconsistency are discussed in [57].

Finally we discuss molecular defects such as CN$^-$ and OH$^-$ substituting for a negative ion in a alkali halide crystal. First consider a cyanide impurity. At first glance, the states of these cigar-shaped polar molecules are quite different from those of a lithium ion. Yet the above model applies equally well to both types of impurities. The tunneling motion of a lithium defect corresponds to the reorientation of the polar molecule between its localized states: the molecular axis is aligned on one of the four space diagonals; together with the two directions for the molecular axis, this leads to eight equivalent positions on the corners of a cube. Hence for cyanide defects, the arrow in Fig. 2.1 corresponds to the orientation of the permanent dipole moment along the C–N axis.

The hydroxyl ion OH$^-$, however, prefers the six orientations along [100] directions, which gives rise to a three-level structure with spacings Δ_0 and $2\Delta_0$, where the tunnel energy takes a value of about $\Delta_0/k_B = 0.35$ K [54].

2.2 Dielectric Response Function

Acoustic and microwave spectroscopy rely on a linear coupling to a weak external field. When tunneling between its eight off-center positions, a lithium ion with charge q carries an electric dipole moment

$$\boldsymbol{p} \equiv q\boldsymbol{r} = \tfrac{1}{2}qd(\alpha,\beta,\gamma) \equiv 3^{-1/2}p(\alpha,\beta,\gamma), \tag{2.17}$$

whose potential energy in an electric field $\boldsymbol{E}(t)$ is given by

$$W = \boldsymbol{E}(t) \cdot \boldsymbol{p}. \tag{2.18}$$

Our notation for the dipole moment (2.17) implies that the absolute value is

$$|\boldsymbol{p}| = p = (\sqrt{3}/2)qd. \tag{2.19}$$

From (2.10), one easily derives that the perturbation W induces only transitions between states whose $\boldsymbol{\sigma}$ vectors differ in one component,

$$|\langle\boldsymbol{\sigma}|\boldsymbol{p}|\boldsymbol{\sigma}'\rangle| = p/\sqrt{3} \qquad \text{for } \tfrac{1}{2}|\boldsymbol{\sigma}-\boldsymbol{\sigma}'|=1, \tag{2.20}$$

where $|\boldsymbol{a}|$ denotes the Eucledian distance $\sqrt{a_x^2+a_y^2+a_z^2}$.

Dynamic dielectric experiments are described in terms of the two-time correlation function of the positon operator (2.2); one easily calculates

$$\tfrac{1}{2}\langle \boldsymbol{r}(t)\boldsymbol{r}(t') + \boldsymbol{r}(t')\boldsymbol{r}(t)\rangle = (d/2)^2 \cos[\Delta_{\rm K}(t-t')/\hbar]. \tag{2.21}$$

In a similar way, we find with $n = N/V$ the dynamic susceptibility of N defects in a sample of volume V

$$\chi_{ij}(t-t') = n(i/\hbar\epsilon_0)\langle[p_i(t), p_j(t')]\rangle\Theta(t-t') \tag{2.22}$$

to be diagonal in i,j and isotropic,

$$\chi_{ij}(t-t') = \frac{2}{3}\frac{np^2}{\hbar\epsilon_0}\tanh(\Delta_{\rm K}/2k_{\rm B}T)\sin[\Delta_{\rm K}(t-t')/\hbar]\delta_{ij}\Theta(t-t'). \tag{2.23}$$

Note that (2.21) and (2.23) are identical to correlation and response functions of a two-level system with energy splitting $\Delta_{\rm K}$.

Experiments provide quantities depending on frequency rather than on time; hence we take the Fourier transform of (2.23),

$$\chi(\omega) = \frac{2}{3}\frac{np^2}{\epsilon_0}\frac{\Delta_{\rm K}}{\Delta_{\rm K}^2 - \hbar^2\omega^2}\tanh(\Delta_{\rm K}/2k_{\rm B}T). \tag{2.24}$$

Separating real and imaginary parts,

$$\chi(\omega) = \chi'(\omega) + i\chi''(\omega), \tag{2.25}$$

the latter or spectral function is given by the discontinuity when crossing the real axis,

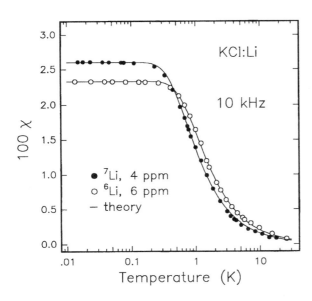

Fig. 2.4. Dielectric susceptibility of Li impurities in KCl:Li in the dilute limit at a frequency $\omega/2\pi = 10$ kHz. The solid lines are given by (2.27) with $^6\Delta_0 = 1.6$ K and $^7\Delta_0 = 1.1$ K; the values for the defect concentration used for the fits read $c = 4.1$ ppm for ^{7}Li and $c = 5.6$ ppm for ^{6}Li. By courtesy of Weis and Enss [62]

$$\chi''(\omega) = \frac{2}{3}\frac{np^2}{\epsilon_0}\left[\pi\delta(\hbar\omega - \Delta_\mathrm{K}) - \pi\delta(\hbar\omega + \Delta_\mathrm{K})\right]\tanh(\Delta_\mathrm{K}/2k_\mathrm{B}T). \quad (2.26)$$

For later use we note the zero-frequency limit of (2.24),

$$\chi = \frac{2}{3}\frac{np^2}{\epsilon_0}\frac{1}{\Delta_\mathrm{K}}\tanh(\Delta_\mathrm{K}/2k_\mathrm{B}T). \quad (2.27)$$

In Fig. 2.4, data measured for 4 ppm ^{7}Li and 6 ppm ^{6}Li are compared with (2.27); note the perfect agreement with respect to temperature dependence and isotope effect.

The tunnel energy Δ_K is larger than $\hbar\omega$ by many orders of magnitudes. Thus here and in the sequel, we neglect the latter term when comparing the real part of the resonant susceptibility with experimental data taken at small but finite frequency.

2.3 Elastic Response Function

Elastic waves are considered for two purposes: first, hybridization of the tunnel states with thermal lattice vibrations leads to a finite width of the energy levels; second, the interaction of the impurities with external acoustic waves results in a change of the sound velocity with temperature.

Because of its misfit in size or shape, the defect at site $\boldsymbol{R}^0 = 0$ causes a local distortion of the lattice, i. e., the surrounding atoms have shifted slightly from the sites $\boldsymbol{R}^m$ which they would occupy in a perfect lattice. Moreover, this shift depends on the precise position of the impurity. Here we are not interested in the overall distortion but rather in the differences arising from various off-center positions, $\boldsymbol{r}$, of the impurity. (A rigorous treatment of the elastic properties of crystals with point defects is given in [33].)

For a lattice distortion that varies sufficiently slowly in space, the interaction potential is given by the term that is linear in the elastic strain $\epsilon_{ij}(\boldsymbol{R})$,

$$W(\boldsymbol{r}) = -Q_{ij}(\boldsymbol{r})\epsilon_{ij}(\boldsymbol{r}). \quad (2.28)$$

Equation (2.28) is the lowest-order term of a multipole expansion with the elastic quadrupole operator

$$Q_{ij}(\boldsymbol{r}) = \sum_n R_j^n K_i^n(\boldsymbol{r}), \quad (2.29)$$

where $\boldsymbol{R}^n$ are the perfect lattice sites of the surrounding crystal atoms and $\boldsymbol{K}^n$ are corresponding Kanzani forces that are related to the actual static shifts through the perfect lattice Green function [33]. For our purpose, it is sufficient to consider $\boldsymbol{K}^n$ as the force exerted by the defect at position $\boldsymbol{r}$ on the atom at site $\boldsymbol{R}^n$, in terms of the atomic potential $\boldsymbol{K}^n(\boldsymbol{r}) = -\nabla V_n(\boldsymbol{R}^n - \boldsymbol{r})$.

We recall that we are looking only for the part of $W(\boldsymbol{r})$ that depends on the defect coordinate $\boldsymbol{r}$. Thus we may expand the force $\boldsymbol{K}^n$ in terms of $\boldsymbol{r}$; with $\partial_i = (\partial/\partial r_i)$ we find

2.3 Elastic Response Function

$$K_i^n(\mathbf{r}) = \partial_i V_n - r_k \partial_i \partial_k V_n + \tfrac{1}{2} r_k r_l \partial_i \partial_k \partial_l V_n + ..., \tag{2.30}$$

where the derivatives of $V_n(\mathbf{R}^n - \mathbf{r})$ are to be evaluated at $\mathbf{r} = 0$. After insertion in (2.29) and using the cubic symmetry of the host crystal, we find for the first term

$$\sum_n R_j^n \partial_i V_n = \text{const.} \times \delta_{ij}. \tag{2.31}$$

The term linear in r_k vanishes because of parity,

$$\sum_n R_j^n \partial_i \partial_k V_n = 0. \tag{2.32}$$

As to the quadratic term, the sum over n is finite only for contributions with pairwise equal labels,

$$\sum_n R_j^n \partial_i \partial_k \partial_l V_n = \frac{4g}{d^2}[\delta_{ik}\delta_{jl} + \delta_{il}\delta_{jk}] + \text{const.} \times \delta_{ij}\delta_{kl}. \tag{2.33}$$

(This equation defines the elastic coupling energy g.)

Noting that the square of the two-state coordinate $r_i^2 = \tfrac{1}{4}d^2$ is a constant, the quadrupole operator reads in quadratic approximation in $\mathbf{r}$,

$$Q_{ij}(\mathbf{r}) = g(2/d)^2 r_i r_j (1 - \delta_{ij}) + \text{const.} \times \delta_{ij}. \tag{2.34}$$

Appropriately symmetrized superpositions of the entries Q_{ij} have been used when dealing with elastic properties of lithium or hydroxyl defects [32,34].

Being constant, the diagonal part Q_{ii} is immaterial for our purpose since it does not mix the tunneling states (2.10). Note that for $i \neq j$ the elastic quadrupole $Q_{ij}(\mathbf{r})$ may be expressed through the T-symmetric representations of O_h [32]. Yet the corresponding radial part of the elastic Green function decays with the third power of distance [33]; for this reason Q_{ij} is sometimes called elastic dipole.

Since the perturbation potential (2.28) is quadratic in the defect coordinates r_i, elastic transitions obey selection rules,

$$|\langle \boldsymbol{\sigma}|\mathbf{Q}(\mathbf{r})|\boldsymbol{\sigma}'\rangle| = g \qquad \text{for } \tfrac{1}{2}|\boldsymbol{\sigma} - \boldsymbol{\sigma}'| = 2, \tag{2.35}$$

which are quite different from those obtained for the electric dipole moment, (2.20); the latter changes a single symmetry label σ_i, whereas (2.35) involves two such labels. According to the spectrum (2.12), electric dipole transitions occur between any pair of states separated by the tunnel energy Δ_K; elastic strain mixes both degenerate states and states with an energy difference of $2\Delta_K$. Since the operator (2.34) is invariant under inversion $\mathbf{r} \to -\mathbf{r}$, it only mixes states of equal parity.

Damping by Thermal Lattice Vibrations

Elastic waves cause a strain field which couples to the local lattice distortion due to the impurity at position r. At low temperatures, the vibrational amplitude

$$u(R) = \sum_k \sqrt{\hbar/2m\omega_k} e(k) e^{ik \cdot R} \left(b_k + b^\dagger_{-k}\right) \tag{2.36}$$

may be considered as slowly varying in space, since only large wave-length phonons are significantly populated. We consider an impurity ion on the lattice site $R^0 = 0$. When we insert the elastic strain

$$\epsilon_{ij}(R) = \partial_j u_i(R) \tag{2.37}$$

in (2.28) and use $r_i = \pm \frac{1}{2}d$, we find the coupling potential

$$W(r) = g \sum_{ij} \sum_k w_{ij}(r) ik_j e_i(k) \sqrt{\hbar/2m\omega_k}(b_k + b^\dagger_{-k}), \tag{2.38}$$

where the quantity w_{ij} embraces the position dependence of both the force operator Q_{ij} and the phase factor $\exp(i k \cdot r)$,

$$\begin{aligned} w_{ij}(r) &= (2/d)^2 \left(r_i \cos(\tfrac{1}{2}k_i d) + \tfrac{1}{2}id \sin(\tfrac{1}{2}k_i d)\right) \\ &\quad \times \left(r_j \cos(\tfrac{1}{2}k_j d) + \tfrac{1}{2}id \sin(\tfrac{1}{2}k_j d)\right)(1 - \delta_{ij}). \end{aligned} \tag{2.39}$$

(Only the off-diagonal elements of the quadrupole tensor $\mathbf{Q}$ have been retained.)

Two types of transitions are mediated by w_{ij}. Besides the quadrupolar term proportional to $r_i r_j$, there is a contribution linear in the coordinates r_i. (An expression similar to (2.39) arises for interstitial hydrogen in Niobium [23].) Note that for large wave-lengths, the latter term is smaller by a factor $\sin(\tfrac{1}{2}k_i d) \approx \tfrac{1}{2}k_i d \ll 1$ than the quadratic one. Hence we discard the linear term and have with $\cos(\tfrac{1}{2}k_i d) \approx 1$ the simple expression

$$w_{ij} = (2/d)^2 r_i r_j (1 - \delta_{ij}), \tag{2.40}$$

which does not depend on the k-vector, in contrast to (2.39).

The time dependence of the vibrational amplitude induces transitions between different tunneling states which acquire a finite life time; the corresponding rates are given by Fermi's Golden Rule

$$\gamma_\sigma = \pi \sum_{k,\sigma'} P_{\sigma \sigma' k} \frac{g^2 \hbar k_j^2}{2d^2 m \omega_k} |\langle \sigma | w_{ij} | \sigma' \rangle|^2 \delta(E_\sigma - E_{\sigma'} \pm \hbar \omega_k), \tag{2.41}$$

where $P_{\sigma \sigma' k}$ is the properly normalized statistical weight involving both phonon and tunneling energies. The delta function assures energy conservation; the two signs of the phonon energy quantum account for upward and downward scattering $E_\sigma < E_{\sigma'}$ and $E_\sigma > E_{\sigma'}$. The actual calculation is simple. For a given initial state σ, a single final level σ' is relevant for phonon scattering; the energy difference is $E_{\boldsymbol{\sigma}} - E_{\boldsymbol{\sigma}'} = \pm 2\Delta_{\rm K}$. (Transitions between degenerate states are insignificant because of the vanishing phonon density of states at $\omega_{\boldsymbol{k}} \to 0$.) Using $\omega_{\boldsymbol{k}} = v|\boldsymbol{k}|$, the definition

$$\gamma_0 = \pi \frac{g^2}{\varrho v^2 \hbar^4 \omega_{\rm D}^3} (2\Delta_{\rm K})^3, \tag{2.42}$$

and the Bose function $n(2\Delta_{\rm K}) = (e^{2\beta \Delta_{\rm K}} - 1)^{-1}$, we find for the widths of the levels in (2.12)

$$\gamma({\rm A}_{1g}) = \gamma({\rm T}_{1u}) = \gamma_0 n(2\Delta_{\rm K}) \equiv \gamma_-, \tag{2.43}$$

$$\gamma({\rm T}_{2g}) = \gamma({\rm A}_{2u}) = \gamma_0 (1 + n(2\Delta_{\rm K})) \equiv \gamma_+. \tag{2.44}$$

We emphasize that w_{ij} as given by (2.40), does not mix states of different parity. Thus relaxation, e.g., from the level $\rm T_{1u}$ to the ground state $\rm A_{1g}$ requires us to take into account the small additional term in (2.39). The rate arising from the contribution linear in r_i is smaller by about five orders of magnitude than γ_0; accordingly, the decay $\rm T_{1u} \to A_{1g}$ occurs very slowly.

The impurity dwelling in one of its off-center positions is accompanied by a specific lattice distortion; when tunneling from one position to another, the ion drags the surrounding atoms into the corresponding state. In terms of the vibrational modes of the crystal, the dressed impurity may be viewed as a small polaron with a renormalized tunneling amplitude. A detailed study for a Li impurity in KCl may be found in [35].

Selection Rules

Sound propagation is probed by an external elastic wave with frequency ω, amplitude A, and polarization vector $\boldsymbol{e}$, which evolves in the direction given by the wave vector $\boldsymbol{k}$. The corresponding strain field reads according to (2.37)

$$\epsilon_{ij}(\boldsymbol{R}) = {\rm i} A(\boldsymbol{R}) k_j e_i. \tag{2.45}$$

Regarding thermal waves, we expand the perturbation potential W in terms of the impurity coordinate $\boldsymbol{r}$ about the site position $\boldsymbol{R}^0$, and find with (2.40) for large wave lengths

$$W(\boldsymbol{r}) = g \sum_{ij} w_{ij}(\boldsymbol{r}) \epsilon_{ij}(\boldsymbol{R}^0). \tag{2.46}$$

Obviously the response of the impurity to an elastic distortion of its neighborhood depends very much on the relative orientation of the vectors $\boldsymbol{k}$ and $\boldsymbol{e}$ with respect to the crystal axes.

As a first example, consider longitudinal waves in [001] directions. Since $\boldsymbol{k}$ and $\boldsymbol{e}$ are parallel, only the diagonal element ϵ_{zz} is finite; with (2.28,2.34) the corresponding potential W is constant and does not mix the tunneling states. Accordingly, there is no linear response to longitudinal waves along a crystal axis.

20   2. Tunneling of a Single Defect

Similar selection rules apply to transverse or shear waves. In Fig. 2.5 we show sound velocity data for KCl with 100 ppm lithium impurities. The shear waves propagate in [110] direction; the sound velocity has been measured for two different polarizations, [001] and [1$\bar{1}$0]. In the first case, [001], there are two finite non-diagonal entries of the strain tensor,

$$\epsilon_{zx} = \epsilon_{zy} = \mathrm{i}2^{-1/2}Ak \equiv \epsilon, \tag{2.47}$$

resulting in a finite value for the perturbation potential,

$$W = g\epsilon(2/d)^2(r_x + r_y)r_z \qquad \text{for [001]}; \tag{2.48}$$

in the second case, [1$\bar{1}$0], the finite entries have the opposite sign,

$$\epsilon_{xy} = -\epsilon_{yx} = \tfrac{1}{2}\mathrm{i}Ak, \tag{2.49}$$

and thus cancel each other out when we calculate the coupling potential,

$$W = 0 \qquad \text{for [1$\bar{1}$0]} . \tag{2.50}$$

Equations (2.48) and (2.50) account for the data shown in Fig. 2.5, where the linear response has been found to vanish for shear waves with polarization direction [1$\bar{1}$0].

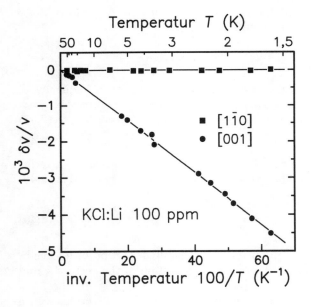

Fig. 2.5. Change of transverse sound velocity arising from 100 ppm Li impurities in KCl, for two different polarizations, [001] and [1$\bar{1}$0]. The shear waves with a frequency of 30 MHz propagate in [110] direction. Regarding the temperature dependence, compare (2.63). Data are from [28]. Drawing by courtesy of Weis [41]

Dynamic Response Function

Linear response theory with respect to elastic strain relies on time correlations of the quadrupole operators w_{ij}; all dynamic information is contained in the symmetrized correlation matrix

$$C_{ijkl}(t - t') = \tfrac{1}{2}\langle w_{ij}(t)w_{kl}(t') + w_{kl}(t')w_{ij}(t)\rangle. \tag{2.51}$$

First we discuss an isolated impurity, i.e., the case of vanishing coupling to the heat bath, and zero damping. Only diagonal elements C_{ijij} and C_{ijji} are of interest. The calculation is rather tedious but straightforward; because of the cubic symmetry, those entries are equal to

$$C^0(\tau) = \left(1 - \tfrac{1}{2}\text{sech}(\beta\Delta_K/2)^2\right)\cos(2\Delta_K\tau/\hbar) + \tfrac{1}{2}\text{sech}(\beta\Delta_K/2)^2. \tag{2.52}$$

(The label '0' indicates zero damping.) Comparison with the position-position correlation (2.21) reveals two basically different features. *First*, (2.52) exhibits both an oscillating and a constant contribution. The amplitude of the latter, $\tfrac{1}{2}\text{sech}(\beta\Delta_K/2)^2$, vanishes at the limit of zero temperature. *Second*, the oscillation frequency $2\Delta_K/\hbar$ is twice as large as that in (2.21).

A glance at Fig. 2.2 on p. 12 renders obvious the origin of both issues; the zero frequency feature in (2.52) arises from transitions between degenerate states of the triplets T_{1u} and T_{2g}.

Now we take into account the coupling to the heat bath. Time evolution of the impurity operators $w_{ij}(t)$ is given by the coordinate $\boldsymbol{r}(t)$ in the Heisenberg picture [44], with the Hamiltonian consisting of the tunneling part, the coupling potential (2.38), and the phonon part $H_{\text{ph}} = \sum_{\boldsymbol{k}} \hbar\omega_{\boldsymbol{k}} b^\dagger_{\boldsymbol{k}} b_{\boldsymbol{k}}$. As a consequence time-correlations vanish in the long-time limit; with $\langle w_{ij}\rangle = 0$ one finds [44,132]

$$\lim_{\tau\to\infty} C_{ijkl}(\tau) = \langle w_{ij}\rangle\langle w_{kl}\rangle = 0. \tag{2.53}$$

The dynamics is governed by the Liouville operator $L = L_{\text{rev}} + L_{\text{irr}}$, whose first part is defined by the Heisenberg equation $L_{\text{rev}}A = [H, A]$ and whose second part accounts for phonon damping. In the relevant temperature range damping is weak, and the phonon coupling may be treated in second-order perturbation theory [161]. As a further approximation, we retain only the diagonal part of L_{irr}, resulting in the disappearance of non-diagonal terms of (2.51) with $i \neq k$ and $j \neq l$. (Strictly speaking, because of the presence of degenerate states, there is no rigorous argument for our neglecting these non-diagonal correlations; cf. [36].)

Proceeding as in [128] or in Chap. 6, we find the finite correlation functions $C_{ijij}(\tau)$ and $C_{ijji}(\tau)$ with $i \neq j$ to be given by

$$\begin{aligned} C(\tau) &= \left(1 - \tfrac{1}{2}\text{sech}(\beta\Delta_K/2)^2\right)\cos(2\Delta_K\tau/\hbar)e^{-\gamma\tau} \\ &\quad + \tfrac{1}{2}\text{sech}(\beta\Delta_K/2)^2 \sum_{\pm} f(\pm\Delta_K)e^{-\gamma_\pm\tau}, \end{aligned} \tag{2.54}$$

where we have used the longitudinal rates $\gamma_\pm$ as defined in (2.43) and (2.44), the transverse rate

$$\gamma = \tfrac{1}{2}(\gamma_+ + \gamma_-) = \tfrac{1}{2}\gamma_0 \coth(\Delta_K/k_B T), \tag{2.55}$$

and the Fermi function $f(\Delta_K) = (1 + e^{\beta \Delta_K})^{-1}$.

In order to study in more detail how the tunneling impurities affect sound propagation, we consider the linear response to the perturbation potential $W = \sum_{ij} Q_{ij} \epsilon_{ij}$, which is described in terms of the elastic response function

$$R_{ijkl}(t - t') = (i/\hbar)\langle[w_{ij}(t), w_{kl}(t')]\rangle \Theta(t - t'). \tag{2.56}$$

In view of the above selection rules and the cubic symmetry of the problem, we restrict ourselves to the diagonal elements $R_{ijij} \equiv R$, which exhibits all features relevant to our purpose.

According to the correlation function (2.54), the elastic susceptibility contains both van Vleck resonance and Debye relaxation contributions. Taking the Fourier transform of $C(\tau)$, and applying the fluctuation dissipation theorem

$$R''(\omega) = (2/\hbar) \tanh(\beta \omega/2) C''(\omega), \tag{2.57}$$

yields the complex elastic response function [61]

$$\begin{aligned} R(\omega) &= \tanh(\beta \Delta_K/2) \frac{4 \Delta_K}{(2 \Delta_K)^2 - \hbar^2 (\omega + i\gamma)^2} \\ &+ \frac{1}{2 k_B T} \operatorname{sech}(\beta \Delta_K/2)^2 \sum_\pm f(\pm \Delta_K) \frac{i \gamma_\pm}{\omega + i \gamma_\pm}. \end{aligned} \tag{2.58}$$

There are several features worth mentioning. The resonances occur at $2\Delta_K$, which is twice the level splitting in (2.12), and the relaxation part shows a complicated temperature dependence with two different rates $\gamma_\pm$. (At low temperature $T \ll \Delta_K/k_B$, the rate γ_- is exponentially small, $\gamma_- = \gamma_0 e^{-2\beta \Delta_K}$, whereas $\gamma_+ = \gamma_0$.)

Sound Velocity

The presence of a few ppm tunneling impurities may significantly affect the sound propagation at low temperatures. As an example, we compare in Fig. 2.6 the sound velocity of pure KCl with that of samples containing 70 ppm ^{6}Li or 60 ppm ^{7}Li defects. Note that up to 10 K the temperature dependence is dominated by the impurities.

The relative change of the sound velocity due to N impurities in a volume V is related to the real part of the elastic response function through

$$\frac{\delta v}{v} = -\frac{n g^2}{2 \varrho v^2} \Re R(\omega), \tag{2.59}$$

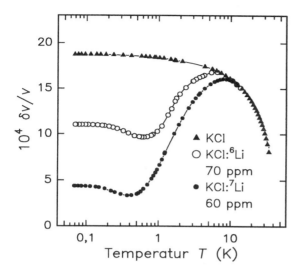

Fig. 2.6. Change of sound velocity of pure KCl and of two samples containing 60 ppm ^{7}Li and 70 ppm ^{6}Li, respectively. For theoretical curves see Fig. 2.7. The data have been observed at a frequency $\omega/2\pi = 2$ kHz. By courtesy of Hübner and Enss [43]

where $n = N/V$ is the impurity density, and ϱ the mass density of the host crystal. For small frequencies $\omega \ll \Delta_\mathrm{K}/\hbar$ we may neglect ω in the first, or resonant, part of (2.58), and thus find

$$\begin{aligned}\frac{\delta v}{v} = &-\frac{1}{2}\frac{ng^2}{\varrho v^2}\left[\frac{1}{\Delta_\mathrm{K}}\tanh\left(\frac{\Delta_\mathrm{K}}{2k_\mathrm{B}T}\right)\right.\\&\left.+\frac{1}{2k_\mathrm{B}T}\mathrm{sech}\left(\frac{\Delta_\mathrm{K}}{2k_\mathrm{B}T}\right)^2 F(T,\omega)\right],\end{aligned} \qquad (2.60)$$

with the weight function of the relaxation contribution

$$F(T,\omega) = \frac{1}{1+e^{-\Delta_\mathrm{K}/k_\mathrm{B}T}}\frac{\gamma_-^2}{\omega^2+\gamma_-^2} + \frac{1}{1+e^{\Delta_\mathrm{K}/k_\mathrm{B}T}}\frac{\gamma_+^2}{\omega^2+\gamma_+^2}. \qquad (2.61)$$

The factor $F(T,\omega)$ results in a strong frequency and temperature dependence of the sound velocity. As a function of frequency, the relaxation term disappears for $\omega \gg \gamma_0$. In the opposite case $\omega \ll \gamma_0$ the sound velocity shows a characteristic minimum as a function of temperature (Fig. 2.7), which hardly depends on frequency, as is obvious when noting

$$F(T,\omega) \to 1 \quad \text{for} \quad \omega/\gamma_0 \to 0. \qquad (2.62)$$

In the intermediate regime $\omega \approx \gamma_0$, the relaxation feature gets weakened and distorted.

In the zero-frequency limit, Eq. (2.62), both the resonant and the relaxation contributions are similar to those of a two-level system with a finite asymmetry energy, and seem to arise from a resonance energy Δ_K. Yet all

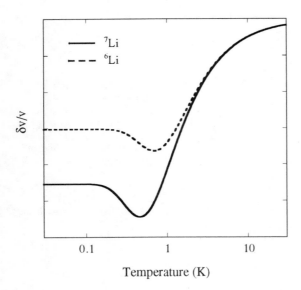

Fig. 2.7. Temperature dependence of the sound velocity for both isotopes at zero frequency. The curves are given by (2.60) with the respective tunnel energies $^7\Delta_K/k_B = 1.1$ K and $^6\Delta_K/k_B = 1.65$ K

physical transitions involve $2\Delta_K$. The temperature dependence of the sound velocity is a subtle result of the product of the factors in (2.52) with that in the fluctuation-dissipation theorem.

In fact, the zero-frequency limit of the sound velocity can be derived in a simpler fashion from the tunneling Hamiltonian supplemented by a constant strain field. Evaluating the entries $\langle\boldsymbol{\sigma}|W|\boldsymbol{\sigma}'\rangle$, diagonalizing the eight-dimensional Hamilton matrix $T+W$, and taking the second derivative with respect to the elastic strain, one gets both the internal energy and the static elastic response function, which, in turn, provides the change in the sound velocity according to (2.59). To our knowledge, the zero-frequency result was first obtained by Hübner [43].

In order to permit a comparison with the data of Fig. 2.6, we plot in Fig. 2.7 the theoretical curve (2.60) with the tunnel energies of both Li isotopes. Note that there is no adjustable parameter except for the scale of the $\delta v/v$ axis, corresponding to the overall factor $g^2/\varrho v^2$. After subtracting the temperature variation due to the host crystal and weighting the curves in Fig. 2.7 with the concentrations 60 ppm for ^{7}Li and 70 ppm for ^{6}Li, one finds the data agree well with (2.60). In Fig. 2.7 we plot the zero-frequency result $F(T,\omega)=1$. Note that the data shown in Fig. 2.6 have been observed at small but finite frequency, $\omega/2\pi = 2$kHz.

We emphasize the different isotope effect of the resonant and the relaxation contributions. The former, proportional to the hyperbolic tangent function, depends on the tunnel energy mainly through the prefactor $1/\Delta_K$, whereas the latter is responsible for the relaxation maximum at about $\Delta_K/2$.

Measuring the frequency dependence of the sound velocity would permit a thorough comparison with the law (2.60). In the range $\omega \approx \gamma_0$ one expects the bump in the curves of Fig. 2.7 to disappear with rising frequency. Such an experiment would permit to determine accurately the elastic coupling energy g.

At high temperatures quantum effects are immaterial, and for both isotopes one finds the classical relation

$$\frac{\delta v}{v} = -\frac{ng^2}{\rho v^2} \frac{1}{2k_B T} \qquad \text{for } T \gg \Delta_K/k_B. \tag{2.63}$$

The data shown in Fig. 2.5 for shear waves with polarization direction [001] obey perfectly well the inverse temperature law of (2.63).

2.4 Thermal Conductivity

In pure crystals, the thermal conductivity κ is determined by phonon scattering from the sample boundaries, leading to the characteristic temperature dependence $\kappa \propto T^3$. The presence of a few substitutional impurities strongly affects the propagation of sound waves, and thus their capacity for transporting heat through the crystal.

When comparing the thermal conductivity of pure KCl with that of samples containing $7 \times 10^{17} \text{cm}^{-3}$ of either ^{6}Li or ^{7}Li, Pohl and his co-workers [30,54] found a significant reduction in the heat transport in the doped samples. Moreover, the data displayed an isotope effect in agreement with specific heat measurements shown in Fig. 2.3.

Scattering off the impurity leads to a finite phase coherence time, or mean free path, for a sound wave, or phonon, traveling through the crystal. The elastic deformation arising from an impurity ion may be decomposed into two parts. The first one is independent of the defect coordinate r, as the constant term in (2.34), whereas the second one is closely related to the motion of the impurity. Only the latter term is relevant for the quantum dynamics of the defect.

According to the selection rules for elastic transition derived above, phonon scattering should occur only between states of the same parity, and it should involve either quasi-elastic transitions or a resonance frequency of $2\Delta_K/\hbar$. Yet from heat conductivity data reported in [30,54,55], it has been conjectured that phonon scattering would involve resonance energies of Δ_K, $2\Delta_K$, and $3\Delta_K$, thus violating the above selection rules.

Deriving the effect of impurity scattering on the heat conductivity is not a simple matter. In order to arrive at an expression permitting a straightforward comparison with measured data, the theoretical model used in [30,54,55] relies on two essential approximations. *First*, any angular dependence of the phonon scattering matrix elements has been discarded. Given the cubic symmetry, this assumption would seem of little consequence. *Second*, the tunneling impurity has been replaced by a classical elastic dipole with resonance frequency ω_0. In view of the complicated temperature dependence of the elastic response function (2.58), the validity of the second approximation is far less obvious.

In terms of the phonon Green function, impurity scattering gives rise to a self-energy whose imaginary part, or phonon damping rate, is given by the imaginary part of the elastic response function, $R''(\omega)$. Yet this function exhibits several features which are not accounted for by the classical dipole model. As discussed on p. 22, $R''(\omega)$ has resonances at $\pm 2\Delta_K$, but with an unusual temperature factor involving an energy Δ_K. Moreover, there is a relaxation term whose temperature dependence is, again, governed by the tunnel energy Δ_K, and not $2\Delta_K$ as the selection rules would suggest.

In order to settle the question about the elastic transitions allowed, it would seem most desirable to incorporate the proper elastic response function $R(\omega)$ in the description of heat transport data.

2.5 Summary

Starting from the tunneling model with eight off-center positions with tunneling between nearest minima only, various static and dynamic properties have been derived. With well-known values for the parameters tunnel energy Δ_K and dipole moment p, the theory accounts for a host of data on specific heat, dielectric response, and sound propagation, concentrating in particular on temperature dependence and isotope effect.

The comparison of (2.24) and (2.58), or of Figs. 2.4 and 2.7, reveals some striking differences between the elastic and the dielectric response of the tunneling impurity. The latter shows no relaxation contribution, and the resonance energies of the former are given by twice the tunnel energy Δ_K. This may be traced back to the symmetry properties of the transition operators. (Compare Fig. 2.2.) The electric dipole moment is odd with respect to parity; hence it mediates transitions between pair and unpair states. On the other hand, the elastic quadrupole operator is even under parity; accordingly, it connects states of the same parity.

As an open question, we consider the temperature dependence of the thermal conductivity. On the basis of a classical dipole model, it has been concluded that the selection rules for elastic transitions, as shown in Fig. 2.2, would be irrelevant for Li impurities. A more thorough investigation of the effect of impurities on heat transport should permit us to determine whether or not the selection rules mentioned are violated .

3. Tunneling of a Coupled Defect Pair

The random configuration of the lithium ions on the host lattice leads to a finite probability of the occurrence of pairs of nearby impurities. In this chapter we deal with pairs of this kind, whose dipolar interaction exceeds the tunnel energy by several orders of magnitude, thus permitting a perturbative approach. Since the resulting energy spectrum and the tunneling dynamics depend very much on the relative orientation of the defects, we treat three particular configurations separately, following closely the original discussion by Weis et al. [40].

The strong dipolar interaction of two nearby defects gives rise to pair excitations well below the bare tunnel energy Δ_K; the spectrum is basically different from that of a single ion. Rotary echoes provide a suitable means for studying impurity pairs on next-nearest-neighbor sites. In order to carry out a thorough comparison with the microscopic model using the available data, we calculate the energy spectrum and the dielectric response function. Particular emphasis is put on the isotope effect with respect to ^{7}Li and ^{6}Li.

3.1 Perturbation Theory

The Hamiltonian consists of three parts which account for the kinetic energy connected to the tunneling motion K, for interaction of the two impurities W, and for an additional one-particle potential V,

$$H = K + V + W. \tag{3.1}$$

According to the previous chapter, each ion has eight possible positions; hence there are 64 product states for the coupled pair,

$$|\alpha\beta\gamma; \alpha'\beta'\gamma'\rangle, \tag{3.2}$$

Greek letters taking the values ± 1.

Kinetic energy

The tunneling part of the Hamiltonian consists of the sum of the kinetic energy of the two defects,

$$K = T + T', \tag{3.3}$$

where T' has the same structure as T; in order to account for an isotope effect arising from a different mass of the second defect, we allow for tunneling amplitudes Δ'_K, Δ'_F, and Δ'_R, as compared with those appearing in (2.9). Obviously T acts on the first three variables in (3.2), and T' on the second triplet.

One-Particle Potential

Residual interactions with other impurities or crystal imperfections break the cubic symmetry of an isolated defect; we thus add a one-particle potential for each of the two defects considered, $V = v(\boldsymbol{r}) + v'(\boldsymbol{r}')$. (The two terms fulfil $v(\boldsymbol{r}) = v'(\boldsymbol{r}' - \boldsymbol{R})$, where the two impurities are separated by $\boldsymbol{R}$.) For sufficiently distant sources, V may be expanded in terms of the the relative positions $\boldsymbol{r}$ and $\boldsymbol{r}'$, and the potential gradients will be the same at both impurity sites, $-\boldsymbol{K} = \nabla v = \nabla' v'$; in linear order we have

$$V = -\boldsymbol{K} \cdot \boldsymbol{r} - \boldsymbol{K} \cdot \boldsymbol{r}', \tag{3.4}$$

which is obviously diagonal in the states in (3.2).

Dipolar Interaction

Because of its net charge q, a lithium impurity carries a bare dipole moment $\boldsymbol{p} = q\boldsymbol{r}$, with an absolute value $(\sqrt{3}/2)qd \approx 2.63\,\text{Debye}$ [31]. Considering two impurities with dipole moments $\boldsymbol{p}$ and $\boldsymbol{p}'$ and distance vector $\boldsymbol{R}$, and expanding the interaction potential in terms of inverse powers of R, the lowest-order approximation yields the dipolar term

$$W = \frac{1}{4\pi\epsilon_0\epsilon}\left(\frac{\boldsymbol{p}\cdot\boldsymbol{p}'}{R^3} - 3\frac{(\boldsymbol{p}\cdot\boldsymbol{R})(\boldsymbol{p}'\cdot\boldsymbol{R})}{R^5}\right). \tag{3.5}$$

This coupling energy is largest for nearby neighbors; the discrete values of the distance vector $\boldsymbol{R}$ and the dipole moments then lead to well separated energy levels. For an ideal host lattice with zero polarizability (i.e., $\epsilon = 1$), the energy scale of (3.5) is given by

$$J_0 = \frac{1}{4\pi\epsilon_0}\frac{q^2d^2}{R^3} = \frac{1}{4\pi\epsilon_0}\frac{4p^2}{3R^3}. \tag{3.6}$$

Yet the screening of the impurity dipoles through the surrounding potassium and chloride ions results in a reduction of the dipolar energy

$$J = \frac{1}{4\pi\epsilon_0\epsilon}\frac{q^2d^2}{R^3} = J_0\frac{1}{\epsilon}. \tag{3.7}$$

This effect is most efficient at sufficiently large distances, where one may use the macroscopic dielectric constant ϵ. For atomic distances, however, the screening is incomplete, and ϵ in (3.7) has to be replaced by some effective value ϵ_{eff}.

Second-Order Approximation

Diagonalization of the Hamiltonian is achieved by solving the eigenvalue equation

$$\det(E - H) = 0. \qquad (3.8)$$

Inserting the numbers given in Table 3.1 yields a coupling energy J well above 100 Kelvin for nearest nearest and next-nearest neighbors. Comparison with typical values for tunneling amplitudes and asymmetry shows that the interaction W is by far the largest energy. Thus we are led to start from the ground states of the interaction potential W and to treat $K + V$ as a small perturbation.

The relevant space is restricted to the ground states shown in Fig. 3.1. We recall that both W and V are diagonal in the localized states and that there are no first-order tunneling transitions between the states of the ground state multiplet. When we define the projection on this subspace P and its complement $Q = 1 - P$, we find for the unperturbed ground-state energy

$$E_0 = PWP. \qquad (3.9)$$

Since $PKP = 0$, the first-order term of the perturbation series involves V only,

$$H^{(1)} = PVP. \qquad (3.10)$$

Since the asymmetry potential V is diagonal, this implies that $PVQ = 0$; hence the second-order term reads

$$H^{(2)} = PKQ \frac{1}{E_0 - W} QKP. \qquad (3.11)$$

Equations (3.9–3.11) provide the second-order approximation for the eigenvalue equation (3.8)

$$\det\left(E - E_0 - H^{(1)} - H^{(2)}\right) = 0; \qquad (3.12)$$

Table 3.1. Parameters for three particular configurations of lithium pairs in potassium chloride. J_0 has been calculated from (3.6). $\boldsymbol{P}$ is the total dipole moment in the ground-state multiplet, and $a = 6.23$ Å is the lattice constant. For more data see [41]

	$\boldsymbol{R}$ / a	R / a	R / Å	J_0 / K	$|\boldsymbol{P}|$ / qd
NN1	$(\frac{1}{2}, \frac{1}{2}, 0)$	$2^{-1/2}$	4.41	975	$\sqrt{2}$
NN2	$(1,0,0)$	1	6.23	276	1
NN3	$(1, \frac{1}{2}, \frac{1}{2})$	$(3/2)^{1/2}$	7.64	188	$\sqrt{3}$

its solution requires us to diagonalize the effective Hamiltonian in a subspace of dimension 4, 8, and 2 for the three configurations shown in Fig. 3.1.

Each particular impurity configuration leads to a different level scheme for W, thus constituting a distinct problem. Fig. 3.1 shows the classical ground states with respect to the dipolar energy (3.5) for three particular defect configurations; in terms of the lattice constant $a = 6.23$ Å, their distance is given by $\sqrt{1/2}\,a$ (NN1), a (NN2), and $\sqrt{3/2}\,a$ (NN3); the number of degenerate states is 4, 8, and 2, respectively.

Regarding the most interesting configuration NN2, degenerate perturbation theory in an eight-dimensional space may be quite a tedious matter. The present case, however, is greatly simplified by the symmetry properties of the Hamiltonian. Since the unperturbed term W is diagonal in the ground states, the perturbation analysis can be done in these states.

The tunneling operators T and T' account for single moves either along the crystal directions x, y, or z, or along the face and space diagonals, with different tunneling amplitudes. From Fig. 3.1 we deduce that passing from one configuration NN2 to another requires a single move of each defect. Regarding the configurations NN1, there are more complicated transitions requiring either two tunneling moves of each defect along the edges, or one move each across the faces of the respective cubes. Regarding configuration NN3, passing from one ground state to another necessitates either three edge tunneling moves of each defect, or one space diagonal move, or some combination.

Equation (2.6) states a hierarchy between the three types of tunneling motion. In order to account for the dominant term, we put $T = T_{\rm K}$ for NN2, $T = T_{\rm K} + T_{\rm F}$ for NN1, and $T = T_{\rm K} + T_{\rm F} + T_{\rm R}$ for NN3. We start with the configuration NN2 which provides the most interesting case in view of exper-

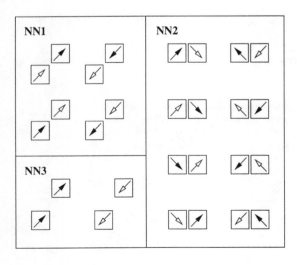

Fig. 3.1. Ground states of two lithium impurities on nearby potassium sites in KCl, schematically. For actual distances see Table 3.1. Arrows indicate the possible dipole orientations projected on the $x - y$ plane. Open arrows denote a negative z component, and filled arrows a positive one. For a discussion of the various configurations see the text. According to [40]

imental restrictions on both frequency and temperature for the observation of rotary echoes.

3.2 Configuration NN2

Dipolar Energy

With the distance vector proportional to the [100] direction, $\boldsymbol{R} = R\boldsymbol{e}_x$, the corresponding dipolar energy reads

$$W = \frac{1}{4\pi\epsilon_0\epsilon}\frac{1}{R^3}\big(-2p_xp'_x + p_yp'_y + p_zp'_z\big). \tag{3.13}$$

When we insert the dipole moments of the two impurities

$$\boldsymbol{p} = 3^{-1/2}p(\alpha,\beta,\gamma), \qquad \boldsymbol{p}' = 3^{-1/2}p(\alpha',\beta',\gamma'),$$

with $p = (\sqrt{3}/2)qd$, it is straightforward to evaluate W in the localized states (3.2). The 64 product states are distributed over 5 levels,

$$\begin{aligned}
E_4 &= & J &: & |\alpha\beta\gamma;\bar{\alpha}\beta\gamma\rangle, \\
E_3 &= & \tfrac{1}{2}J &: & |\alpha\beta\gamma;\bar{\alpha}\bar{\beta}\gamma\rangle, |\alpha\beta\gamma;\bar{\alpha}\beta\bar{\gamma}\rangle, \\
E_2 &= & 0 &: & |\alpha\beta\gamma;\bar{\alpha}\bar{\beta}\bar{\gamma}\rangle, |\alpha\beta\gamma;\alpha\beta\gamma\rangle, \\
E_1 &= & -\tfrac{1}{2}J &: & |\alpha\beta\gamma;\alpha\bar{\beta}\gamma\rangle, |\alpha\beta\gamma;\alpha\beta\bar{\gamma}\rangle, \\
E_0 &= & -J &: & |\alpha\beta\gamma;\alpha\bar{\beta}\bar{\gamma}\rangle,
\end{aligned} \tag{3.14}$$

whose degeneracies read 8:16:16:16:8. The eight ground states, with energy E_0, are drawn in Fig. 3.1.

Ground-State Splitting

Now we turn to the perturbation analysis with respect to V and K. With the short-hand notation for the ground states

$$|G\alpha\beta\gamma\rangle = |\alpha\beta\gamma;\alpha\bar{\beta}\bar{\gamma}\rangle, \tag{3.15}$$

the corresponding projection operators read

$$P = \frac{1}{8}\sum_{\alpha\beta\gamma}|G\alpha\beta\gamma\rangle\langle G\alpha\beta\gamma| \tag{3.16}$$

and $Q = 1 - P$.

Since both W and V are diagonal, it is straightforward to calculate the unperturbed ground-state energy

$$E_0 = PWP = -J \tag{3.17}$$

and the first-order term PVP. Evaluating (3.4) in terms of (3.15) one finds that only the x component yields a finite bias,

3. Tunneling of a Coupled Defect Pair

$$\langle G\alpha\beta\gamma|V|G\kappa\lambda\mu\rangle = \tfrac{1}{2}\Delta\delta_{\alpha\kappa}\delta_{\beta\lambda}\delta_{\gamma\mu}(\delta_{\kappa 1} - \delta_{\kappa\bar{1}}). \tag{3.18}$$

Note that the asymmetry energy of each impurity is given by

$$\tfrac{1}{2}\Delta \equiv K_x d; \tag{3.19}$$

the quantity Δ in (3.18) is the sum of both terms.

Now we turn to the second-order term $H^{(2)}$. For reasons to be discussed below we retain only the dominant term of the tunneling Hamiltonian and thus put $T = T_K$ and $T' = T'_K$. It turns out to be convenient to split $H^{(2)}$ into two parts,

$$H^{(2)} = H^{(2)}_d + H^{(2)}_{nd}, \tag{3.20}$$

where the first one contains the diagonal terms and the second one the remainder. The particular structure of the tunneling Hamiltonian and of the ground states results in the simple relations

$$H^{(2)}_d = T\frac{1}{E_0 - W}T + T'\frac{1}{E_0 - W}T', \tag{3.21}$$

$$H^{(2)}_{nd} = T\frac{1}{E_0 - W}T' + T'\frac{1}{E_0 - W}T. \tag{3.22}$$

[More rigorously, the right-hand side should be multiplied from both sides with P, since $H^{(2)}$ is defined on the subspace spanned by (3.16) only.]

The first term (3.21) describes tunneling of one particle from one position to another and back to the original one. The non-diagonal part implies one tunneling movement of each impurity; from (3.2) it is clear that $H^{(2)}_{nd}$ is finite only between states differing in one label, i.e., that both particles move along the same axis.

We begin with the non-diagonal part $H^{(2)}_{nd}$. First consider the case where the tunneling motion occurs in the x direction, corresponding to a change of the label α in (3.2). Inserting (3.2) and (3.22) we find

$$\langle G\alpha\beta\gamma|H^{(2)}_{nd}|G\bar{\alpha}\beta\gamma\rangle = \frac{\Delta_K \Delta'_K}{4}\frac{2}{E_0 - E_2};$$

the factor 2 arises from the two channels in (3.22), which both pass through an intermediate state with energy E_2.

Transitions involving the third label correspond in real space to one defect moving in the positive the other in the negative z direction. Proceeding as above we find

$$\langle G\alpha\beta\gamma|H^{(2)}_{nd}|G\alpha\beta\bar{\gamma}\rangle = \frac{\Delta_K \Delta'_K}{4}\frac{2}{E_0 - E_1};$$

note that these transitions involve intermediate states with energy E_1. Owing to the symmetry about the x axis, the same result occurs for tunneling in the y direction. Matrix elements involving more than one label vanish in the present second-order approximation with respect to K.

3.2 Configuration NN2

Now we turn to the diagonal contribution (3.21). Since the initial and final states are identical, the three terms for tunneling along the different axes have to be added. Proceeding as above we find

$$\langle G\alpha\beta\gamma|H_d^{(2)}|G\alpha\beta\gamma\rangle = \frac{\Delta_K^2 + \Delta_K'^2}{4}\left(\frac{2}{E_0 - E_1} + \frac{1}{E_0 - E_2}\right).$$

When we put together the above results and the definitions

$$\eta_0 = \frac{\Delta_K^2 + \Delta_K'^2}{2J}, \tag{3.23}$$

$$\eta = \frac{\Delta_K \Delta_K'}{J}, \tag{3.24}$$

we find the second-order term of the perturbation series

$$\langle G\alpha\beta\gamma|H^{(2)}|G\kappa\lambda\mu\rangle = -\tfrac{5}{2}\eta_0 \delta_{\alpha\kappa}\delta_{\beta\lambda}\delta_{\gamma\mu}$$
$$- \eta\left(\tfrac{1}{2}\delta_{\alpha\bar\kappa}\delta_{\beta\lambda}\delta_{\gamma\mu} + \delta_{\alpha\kappa}\delta_{\beta\bar\lambda}\delta_{\gamma\mu} + \delta_{\alpha\kappa}\delta_{\beta\lambda}\delta_{\gamma\bar\mu}\right), \tag{3.25}$$

where, as usual, Greek letters take the values ± 1 and the bar changes the sign, $\bar\alpha = -\alpha$.

Diagonalizing the 8-dimensional matrices (3.18,3.25) permits us to calculate eigenvalues and eigenstates of the effective Hamiltonian

$$\tilde H = E_0 + H^{(1)} + H^{(2)}. \tag{3.26}$$

Owing to the particular structure of $\tilde H$, this diagonalization reduces to three two-dimensional problems. More precisely, $\tilde H$ consists of three terms, each of which involves either x or y or z coordinates; for the eigenstates this means that they factorize with respect to the three labels of the ground states $|G\alpha\beta\gamma\rangle$.

First we consider the motion perpendicular to the x axis, which involves only the second-order term (3.25). Partial diagonalization with respect to the second labels β and λ yields even or odd superpositions of the respective ground states; the same argument holds for the third labels γ and μ. It turns out to be convenient to define new states

$$|G\alpha_x;\sigma_y\sigma_z\rangle = \frac{1}{2}\sum_{\alpha_y,\alpha_z=\pm 1} f_{\alpha_y}^{\sigma_y} f_{\alpha_z}^{\sigma_z}|G\alpha_x\alpha_y\alpha_z\rangle, \tag{3.27}$$

which are diagonal with respect to the last two labels. [The phase factors f_α^σ are defined in (2.11).] Here and in the following text, α_i denote reduced coordinates and σ_i denote quantum numbers of energy eigenstates.

The remaining two-dimensional problem involves terms of both first and second order. Using (3.18) and (3.25), we find the diagonal matrix elements

$$\langle G\alpha_x;\sigma_y\sigma_z|\tilde H|G\alpha_x;\sigma_y\sigma_z\rangle = E_0 - \tfrac{5}{2}\eta_0 + \eta(\sigma_y + \sigma_z) - \tfrac{1}{2}\alpha_x\Delta, \tag{3.28}$$

and the off-diagonal ones

$$\langle G\alpha_x;\sigma_y\sigma_z|\tilde H|G\bar\alpha_x;\sigma_y\sigma_z\rangle = -\tfrac{1}{2}\eta. \tag{3.29}$$

The resulting eigenvalue equation involves a 2×2 matrix in terms of the coordinate α_x; different values for the quantum numbers σ_y and σ_z lead merely to an overall shift of the energies.

Diagonalizing (3.28,3.29) yields the energy spectrum in terms of the quantum numbers $\boldsymbol{\sigma}$,

$$E_{\boldsymbol{\sigma}} = E_0 - \tfrac{5}{2}\eta_0 + \tfrac{1}{2}\sigma_x\sqrt{\eta^2 + \Delta^2} + (\sigma_y + \sigma_z)\eta. \qquad (3.30)$$

For zero asymmetry $\Delta = 0$, the eight states are distributed over six equidistant levels with spacing η. In the case of finite Δ, the energy of the levels $\sigma_x = -1$ is lowered, and that of $\sigma_x = +1$ increases by the same amount. (Compare Fig. 3.2.)

Solving the two-dimensional linear equation (3.28,3.29), we obtain the corresponding eigenstates

$$\frac{1}{\sqrt{2}} \sum_{\alpha_x = \pm 1} \tilde{f}_{\alpha_x}^{\sigma_x} |G\alpha_x; \sigma_y\sigma_z\rangle,$$

where the coefficients are given by

$$\tilde{f}_1^{\sigma_x}(\eta) = \frac{\Delta - \sigma_x\sqrt{\eta^2 + \Delta^2}}{\sqrt{\Delta^2 + \eta^2 - \sigma_x\Delta\sqrt{\eta^2 + \Delta^2}}}, \qquad (3.31)$$

$$\tilde{f}_{-1}^{\sigma_x}(\eta) = \frac{\eta}{\sqrt{\Delta^2 + \eta^2 - \sigma_x\Delta\sqrt{\eta^2 + \Delta^2}}}. \qquad (3.32)$$

By inserting (3.27), we finally obtain the eigenstates in terms of the basis (3.2),

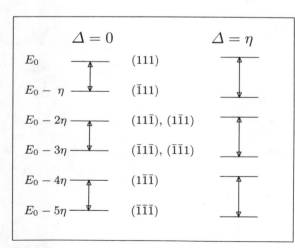

Fig. 3.2. Ground state-splitting for a pair of Li impurities on the x axis, according to (3.30) with $\eta_0 = \eta$. Both cases of zero and finite asymmetry Δ are shown. The states are labelled by the quantum numbers $\boldsymbol{\sigma}$. The arrows indicate the allowed dipole transitions

$$|\boldsymbol{\sigma}\rangle = \frac{1}{\sqrt{8}} \sum_{\{\alpha_i=\pm\}} \tilde{f}_{\alpha_x}^{\sigma_x} \tilde{f}_{\alpha_y}^{\sigma_y} \tilde{f}_{\alpha_z}^{\sigma_z} |G\alpha_x\alpha_y\alpha_z\rangle. \tag{3.33}$$

In the limit $\Delta \to 0$, the coefficients $\tilde{f}_\alpha^\sigma$ tend towards f_α^σ as given by (2.11). In the opposite case $\eta/\Delta \to 0$, we find $f_{\mp 1}^{\pm 1} \to \sqrt{2}$ and $f_{\pm 1}^{\pm 1} \to 0$, which corresponds to states where the two impurities are localized at either $\alpha_x = 1 = \alpha'_x$ or $\alpha_x = -1 = \alpha'_x$.

Internal Energy

From the energy spectrum calculated in the previous section, one easily obtains the statistical operator

$$\rho = \frac{1}{Z} \sum_{\boldsymbol{\sigma}} e^{-E_{\boldsymbol{\sigma}}/k_B T} |\boldsymbol{\sigma}\rangle\langle\boldsymbol{\sigma}|, \tag{3.34}$$

with the partition function $Z = \sum_{\boldsymbol{\sigma}} e^{-E_{\boldsymbol{\sigma}}/k_B T}$ and temperature T. Static properties, such as the specific heat, are determined by the internal energy $U = \langle \tilde{H} \rangle$,

$$U = \tilde{E}_0 - 2\eta \tanh\left(\frac{\eta}{k_B T}\right) - \frac{1}{2}\sqrt{\eta^2 + \Delta^2} \tanh\left(\frac{\sqrt{\eta^2 + \Delta^2}}{2k_B T}\right), \tag{3.35}$$

where $\tilde{E}_0 = E_0 - \frac{5}{2}\eta_0$ and $\langle ... \rangle$ denotes the thermal average $\mathrm{tr}(\rho...)$. (The system is confined to the ground-state multiplet.)

Dielectric Susceptibility

Dynamic quantities are described by the time-dependent response function with respect to an external field $\boldsymbol{F}$,

$$\alpha(t - t') = (i/\hbar\epsilon_0)\langle[\boldsymbol{P}(t), \boldsymbol{P}(t')]\rangle \Theta(t - t'), \tag{3.36}$$

with the total dipole moment of the two impurities

$$\boldsymbol{P} = \boldsymbol{p} + \boldsymbol{p}'. \tag{3.37}$$

Equation (3.36) is to be read as a dyadic product; thus the response function α is a second-rank tensor.

We start by calculating the matrix elements of $\boldsymbol{P}$ in the basis (3.2),

$$\langle G\alpha_x\alpha_y\alpha_z|\boldsymbol{P}|G\alpha'_x\alpha'_y\alpha'_z\rangle = qde_x\ \alpha_x \delta_{\alpha_x\alpha'_x}\delta_{\alpha_y\alpha'_y}\delta_{\alpha_z\alpha'_z}; \tag{3.38}$$

as expected, $\boldsymbol{P}$ is diagonal, and its only non-zero component is along the x axis $\boldsymbol{e}_x$.

Now it is straightforward to derive the matrix elements with respect to the energy eigenstates (3.33),

$$\langle\boldsymbol{\sigma}|\boldsymbol{P}|\boldsymbol{\sigma}'\rangle = \tfrac{1}{2}qde_x \left(\tilde{f}_1^{\sigma_x} \tilde{f}_1^{\sigma'_x} - \tilde{f}_{-1}^{\sigma_x}\tilde{f}_{-1}^{\sigma'_x}\right) \delta_{\sigma_y\sigma'_y}\delta_{\sigma_z\sigma'_z}. \tag{3.39}$$

3. Tunneling of a Coupled Defect Pair

As an important selection rule, we note that dipolar transitions occur only between states with the same labels σ_y and σ_z, i.e., only the x component of the external field induces transitions. As illustrated in Fig. 3.2, all allowed dipole transitions involve the same energy $\sqrt{\eta^2 + \Delta^2}$.

With (3.31,3.32) and $\sigma_x = \sigma'_x$ we use some algebra to find the diagonal elements

$$\frac{1}{2}\left(\tilde{f}_1^{\sigma_x}\tilde{f}_1^{\sigma_x} - \tilde{f}_{-1}^{\sigma_x}\tilde{f}_{-1}^{\sigma_x}\right) = -\sigma_x \frac{\Delta}{\sqrt{\eta^2 + \Delta^2}}. \qquad (3.40)$$

Inserting (3.35) in (3.34) yields the static part of the dipole moment

$$\langle \boldsymbol{\sigma}|\boldsymbol{P}|\boldsymbol{\sigma}\rangle = -\sigma_x\, qd \frac{\Delta}{\sqrt{\eta^2 + \Delta^2}} \boldsymbol{e}_x, \qquad (3.41)$$

which vanishes for zero asymmetry energy and takes opposite values in the states $\sigma_x = \pm 1$. The static part (3.41) leads to a constant term in the correlation function $\langle \boldsymbol{P}(t)\boldsymbol{P}(t')\rangle$; yet in the absence of damping, it does not contribute to the response function (3.31). (If we took phonon coupling into account, it would give rise to a relaxation contribution.)

Non-diagonal transitions with $\boldsymbol{\sigma} \neq \boldsymbol{\sigma}'$ involve the time-dependent part of the dipole moment. When we take $\sigma'_x = -\sigma_x = \bar{\sigma}_x$, we obtain

$$\frac{1}{2}\left(\tilde{f}_1^{\sigma_x}\tilde{f}_1^{\bar{\sigma}_x} - \tilde{f}_{-1}^{\sigma_x}\tilde{f}_{-1}^{\bar{\sigma}_x}\right) = -\frac{\eta}{\sqrt{\eta^2 + \Delta^2}}, \qquad (3.42)$$

and the amplitudes for non-diagonal dipole transitions

$$\langle \sigma_x\sigma_y\sigma_z|\boldsymbol{P}|\bar{\sigma}_x\sigma_y\sigma_z\rangle = -qd \frac{\eta}{\sqrt{\eta^2 + \Delta^2}} \boldsymbol{e}_x. \qquad (3.43)$$

With (3.41) and (3.43), the absolute value of the total dipole moment is found to be

$$|\boldsymbol{P}| = qd. \qquad (3.44)$$

Now we are prepared to evaluate the response function (3.36). The thermal average $\langle ...\rangle$ factorizes with respect to the quantum numbers σ_x, σ_y, and σ_z. Since according to (3.41,3.43) the dipole operator involves σ_x only, we may perform the trace over the variables σ_y and σ_z in (3.36) first.

Thus calculation of the response function reduces to a two-level problem with the states $\sigma_x = \pm 1$. Both the statistical operator and the time evolution operator are diagonal in the energy eigenstates. When we insert

$$\boldsymbol{P}(t) = e^{i\tilde{H}t/\hbar}\boldsymbol{P}e^{-i\tilde{H}t/\hbar}$$

and the spectral representation

$$e^{\pm i\tilde{H}t/\hbar} = \sum_{\boldsymbol{\sigma}} e^{\pm iE_{\boldsymbol{\sigma}} t/\hbar}|\boldsymbol{\sigma}\rangle\langle\boldsymbol{\sigma}|$$

in (3.31), and use the orthogonality relation $\langle\boldsymbol{\sigma}|\boldsymbol{\sigma}'\rangle = \delta_{\boldsymbol{\sigma}\boldsymbol{\sigma}'}$, we obtain, with (3.34,3.42) and

$$E = \sqrt{\eta^2 + \Delta^2}, \tag{3.45}$$

the response function

$$\alpha_{xx}(\tau) = \frac{2}{\hbar\epsilon_0}(qd)^2 \frac{\eta^2}{E^2} \tanh\left(\frac{E}{2k_BT}\right) \sin\left(\frac{E\tau}{\hbar}\right) \Theta(\tau). \tag{3.46}$$

Note that the susceptibility involves only the x direction; there is no linear response along the axes y and z.

Up to now we have dealt with a pair of two impurities on next-nearest-neighbor sites on the x axis. Clearly, if we considered NN2 pairs along the remaining axes y and z, we would find response functions α_{yy} and α_{zz} identical to (3.46).

3.3 Configuration NN1

Now we turn to nearest-neighbor pairs NN1 as shown in Fig. 3.1. Inserting the distance vector $\boldsymbol{R} = 2^{-1/2}R(\boldsymbol{e}_x + \boldsymbol{e}_y)$ in the interaction potential (3.5), we find the 64 states to be grouped in six levels according to

$$\begin{aligned}
E_5 &= \tfrac{5}{4}J &&: |\alpha\alpha\gamma;\bar{\alpha}\bar{\alpha}\gamma\rangle, \\
E_4 &= \tfrac{3}{4}J &&: |\alpha\alpha\gamma;\bar{\alpha}\bar{\alpha}\bar{\gamma}\rangle, |\alpha\bar{\alpha}\gamma;\alpha\bar{\alpha}\gamma\rangle, \\
E_3 &= \tfrac{1}{4}J &&: |\alpha\alpha\gamma;\alpha\bar{\alpha}\gamma\rangle, |\alpha\alpha\gamma;\bar{\alpha}\alpha\gamma\rangle, |\alpha\bar{\alpha}\gamma;\alpha\bar{\alpha}\bar{\gamma}\rangle, \\
E_2 &= -\tfrac{1}{4}J &&: |\alpha\alpha\gamma;\alpha\bar{\alpha}\bar{\gamma}\rangle, |\alpha\alpha\gamma;\bar{\alpha}\alpha\bar{\gamma}\rangle, |\alpha\bar{\alpha}\gamma;\bar{\alpha}\alpha\gamma\rangle, \\
E_1 &= -\tfrac{3}{4}J &&: |\alpha\alpha\gamma;\alpha\alpha\gamma\rangle, |\alpha\bar{\alpha}\gamma;\bar{\alpha}\alpha\bar{\gamma}\rangle, \\
E_0 &= -\tfrac{5}{4}J &&: |\alpha\alpha\gamma;\alpha\alpha\bar{\gamma}\rangle.
\end{aligned} \tag{3.47}$$

The degeneracies 4:8:20:20:8:4 are obtained by putting α and γ independently to ± 1 and by exchanging the first and the second coordinate triplet.

The four ground states with energy $E_0 = PWP = -\tfrac{5}{4}J$ are denoted by

$$|G\alpha\gamma\rangle = |\alpha\alpha\gamma;\alpha\alpha\bar{\gamma}\rangle \tag{3.48}$$

and define the projection operators $P = \tfrac{1}{2}\sum_{\alpha\gamma}|G\alpha\gamma\rangle\langle G\alpha\gamma|$ and $Q = 1 - P$. When we take $\Delta = 2(K_x + K_y)d$, we find as finite matrix elements of the asymmetry potential

$$\langle G\alpha\gamma|V|G\alpha\gamma\rangle = \tfrac{1}{2}\alpha\Delta. \tag{3.49}$$

In order to obtain eigenstates involving all four localized configurations NN1, we need to take into account face diagonal tunneling T_F besides the dominant term T_K, contrary to the situation encountered for NN2. When we take

$$T = T_\mathrm{K} + T_\mathrm{F} \tag{3.50}$$

and T' accordingly, and separate diagonal and non-diagonal parts of the second-order term (3.11), we find the former to contain both T_K and T_F,

$$\langle G\alpha\gamma|H^{(2)}|G\alpha\gamma\rangle = -\frac{\Delta_K^2 + \Delta_K'^2}{J} - \frac{11}{24}\frac{\Delta_F^2 + \Delta_F'^2}{J} \equiv -2\eta_0 - \xi_0. \tag{3.51}$$

The first term on the right-hand side arises from intermediate levels E_1 and E_2, the second one from E_3 and E_4.

As to the off-diagonal matrix elements, there are two types involving either T_K and T_K' or T_F and T_F',

$$\langle G\alpha\gamma|H^{(2)}|G\alpha\bar{\gamma}\rangle = \frac{1}{2}\frac{\Delta_K \Delta_K'}{E_0 - E_1} = -\eta, \tag{3.52}$$

$$\langle G\alpha\gamma|H^{(2)}|G\bar{\alpha}\gamma\rangle = \frac{1}{2}\frac{\Delta_F \Delta_F'}{E_0 - E_4} = -\xi, \tag{3.53}$$

with respective intermediate levels E_4 and E_1.

Diagonalization of (3.49–3.53) runs along the same lines as in the case NN2; the four-dimensional matrix breaks up in a pair of two-dimensional matrices. Accordingly, the four levels may be written by means of two quantum numbers σ and σ_z,

$$E_{\sigma\sigma_z} = E_0 - 2\eta_0 - \xi_0 + \frac{1}{2}\sigma\sqrt{\xi^2 + \Delta^2} + \sigma_z\eta. \tag{3.54}$$

Similarly, for the eigenstates we derive

$$|\sigma\sigma_z\rangle = \frac{1}{2}\sum_{\alpha\gamma} \tilde{f}_\alpha^\sigma f_\gamma^{\sigma_z}|G\alpha\gamma\rangle, \tag{3.55}$$

where f_α^σ is given by (2.11) and $\tilde{f}_\alpha^\sigma = \tilde{f}_\alpha^\sigma(\xi)$ by (3.31,3.32) evaluated at ξ.

When we proceed as for NN2, both static and dynamic quantities are easily derived from the eigenstates (3.55); here we merely note the main results. For the internal energy we find with $\tilde{E}_0 = E_0 - 2\eta_0 - \xi_0$

$$U = \tilde{E}_0 - \eta\tanh\left(\frac{\eta}{k_B T}\right) - \frac{1}{2}\sqrt{\xi^2 + \Delta^2}\tanh\left(\frac{\sqrt{\xi^2 + \Delta^2}}{2k_B T}\right). \tag{3.56}$$

In the ground state multiplet, the total dipole moment $\boldsymbol{P} = \boldsymbol{p} + \boldsymbol{p}'$ is parallel to the distance vector $\boldsymbol{R}$,

$$\langle \sigma\sigma_z|\boldsymbol{P}|\sigma'\sigma_z'\rangle = \frac{1}{2}qd\left(\tilde{f}_1^\sigma \tilde{f}_1^{\sigma'} - \tilde{f}_{-1}^\sigma \tilde{f}_{-1}^{\sigma'}\right)\delta_{\sigma_z\sigma_z'}(\boldsymbol{e}_x + \boldsymbol{e}_y); \tag{3.57}$$

inserting (3.40,3.42) yields the same expressions as above, but with $\boldsymbol{e}_x$ replaced by $(\boldsymbol{e}_x + \boldsymbol{e}_y)$. Accordingly, the absolute value is larger than (3.44):

$$|\boldsymbol{P}| = \sqrt{2}qd. \tag{3.58}$$

The energy spectrum for configuration NN1 is distinct from that obtained for NN2. The most significant difference arises for the dynamical susceptibility

$$\alpha_{ij}(\tau) = \frac{2}{\hbar\epsilon_0}(qd)^2\frac{\xi^2}{\xi^2 + \Delta^2}\tanh\left(\frac{\sqrt{\xi^2 + \Delta^2}}{2k_B T}\right)\sin\left(\frac{\sqrt{\xi^2 + \Delta^2}\,\tau}{\hbar}\right) \tag{3.59}$$

for $i, j = x, y$. Whereas the spectrum exhibits both energies η and $\sqrt{\xi^2 + \Delta^2}$, the response function involves only the latter. For relevant values of the asymmetry one has $\sqrt{\xi^2 + \Delta^2} \ll \eta$.

Contrary to (3.46), here the response function is finite in the $x - y$ plane, i.e., for i and j equal x, y. For a configuration described by the distance vector $\boldsymbol{R} = 2^{-1/2} R(\boldsymbol{e}_x - \boldsymbol{e}_y)$ the non-diagonal elements of α_{ij} would carry a minus sign; averaging over all possible configurations thus cancels the non-diagonal part of (3.59).

3.4 Configuration NN3

Finally we consider two impurities on adjacent sites along a $[1\frac{1}{2}\frac{1}{2}]$ crystal direction. With the distance vector $\boldsymbol{R} = \sqrt{2/3} R(1, \frac{1}{2}, \frac{1}{2})$, the dipolar interaction (3.5) yields a six-level scheme

$$E_5 = \tfrac{5}{4}J \ : |\alpha\alpha\alpha; \bar{\alpha}\bar{\alpha}\bar{\alpha}\rangle,$$
$$E_4 = \tfrac{3}{4}J, \quad E_3 = \tfrac{1}{4}J, \quad E_2 = -\tfrac{1}{4}J, \quad E_1 = -\tfrac{3}{4}J,$$
$$E_0 = -\tfrac{5}{4}J \ : |\alpha\alpha\alpha; \alpha\alpha\alpha\rangle, \tag{3.60}$$

with degeneracies 2:14:16:16:14:2. (We have indicated explicitly only the states which we need.) Space diagonal tunneling T_R is the only part of (2.8) which contributes in second order off-diagonal matrix elements to the effective Hamiltonian $\tilde{H}$; thus we use the full tunneling operator

$$T = T_K + T_F + T_R. \tag{3.61}$$

There are only two ground states $|G\alpha\rangle = |\alpha\alpha\alpha; \alpha\alpha\alpha\rangle$ with the projection $P = \sum_\alpha |G\alpha\rangle\langle G\alpha| = 1 - Q$. When we absorb the irrelevant term $\langle G\alpha|H^{(2)}|G\alpha\rangle$ in the ground state energy $\tilde{E}_0$, the effective Hamiltonian is determined by the diagonal matrix element

$$\langle G\alpha|V|G\alpha\rangle = \tfrac{1}{2}\alpha\Delta \tag{3.62}$$

and the off-diagonal one

$$\langle G\alpha|H^{(2)}|G\bar{\alpha}\rangle = \frac{1}{2}\frac{\Delta_R \Delta'_R}{E_0 - E_5} = -\frac{1}{5}\frac{\Delta_R \Delta'_R}{J} \equiv -\frac{1}{2}\zeta. \tag{3.63}$$

Diagonalization yields the eigenvalues

$$E_\sigma = \tilde{E}_0 + \tfrac{1}{2}\sigma\sqrt{\zeta^2 + \Delta^2} \tag{3.64}$$

and the corresponding eigenstates

$$|\sigma\rangle = \frac{1}{\sqrt{2}} \sum_\alpha \tilde{f}^\sigma_\alpha |G\alpha\rangle, \tag{3.65}$$

where $\tilde{f}^\sigma_\alpha = \tilde{f}^\sigma_\alpha(\zeta)$. The dipole moment is similar to the expressions derived above,

$$\langle \sigma | \boldsymbol{P} | \sigma' \rangle = \frac{1}{2} q d \left(\tilde{f}^\sigma_1 \tilde{f}^{\sigma'}_1 - \tilde{f}^\sigma_{-1} \tilde{f}^{\sigma'}_{-1} \right) (\boldsymbol{e}_x + \boldsymbol{e}_y + \boldsymbol{e}_z); \tag{3.66}$$

yet its absolute value exceeds those of the configurations NN2 and NN3:

$$|\boldsymbol{P}| = \sqrt{3} q d. \tag{3.67}$$

The dynamical response function

$$\alpha_{ij}(\tau) = \frac{2}{\hbar \epsilon_0} (qd)^2 \frac{\zeta^2}{\zeta^2 + \Delta^2} \tanh\left(\frac{\sqrt{\zeta^2 + \Delta^2}}{2 k_B T} \right) \sin\left(\frac{\sqrt{\zeta^2 + \Delta^2} \tau}{\hbar} \right) \tag{3.68}$$

(for $\tau \geq 0$) consists of a matrix with nine identical entries α_{ij} with $i,j = x, y, z$. When we average over the various configurations NN3 (e.g., $\boldsymbol{R} = \sqrt{2/3} R(\frac{1}{2}, -\frac{1}{2}, -1)$), the non-diagonal entries of α_{ij} vanish, and (3.68) has to be multiplied by δ_{ij}.

3.5 Rotary Echoes

First we will apply the preceeding theory on rotary echoes observed for KCl:Li; then we will discuss data measured for three samples with 60 ppm ^{7}Li, 70 ppm ^{6}Li, and 33 ppm of both ^{6}Li and ^{7}Li. Since the total impurity concentration lies between 60 and 70 ppm in all cases, a thorough comparison can be made with respect to the variation of the isotope mass.

Two-State Approximation

As a most surprising result of the above theory, the dynamical response function is formally identical to that of a *single* two-state system. This statement holds true for all three configurations investigated, with respective 'tunnel energies' η, ξ, and ζ, and response functions given by (3.46) for NN2, (3.59) for NN1, and (3.68) for NN3. We stress that this coincidence is not a trivial matter; as an essential condition, we note the particular form of the asymmetry potential V, (3.4).

As a consequence, in linear-response approximation a coupled impurity pair may be substituted by a two-level system with energy splitting E and non-diagonal dipole moment (3.43), for the case NN2; this holds true for the non-linear response as long as the potential energy is smaller than the effective tunnel splitting, $|\boldsymbol{P} \cdot \boldsymbol{F}| \ll \eta$ (for NN2). Thus we may use the well-known theory for rotary echoes of two-level systems [83–85,166], which will greatly simplify the discussion.

Echo experiments do not give the average of all impurities present in a given sample but single out those satisfying the resonance condition $E = \hbar \omega$, where E is the relevant energy and $\omega/2\pi$ the frequency of the external field. For frequencies of about 1 GHz, only next-nearest-neighbor pairs of the type NN2 are likely to fulfil that condition. Thus our discussion relies on the corresponding results of Sect. 3.2.

The distance between the impurities of a pair NN2 is equal to the lattice constant, corresponding to a fixed value for the dipolar interaction J, and with (3.24) to a fixed tunnel splitting η. According to (3.37), the resonance condition $E = \hbar\omega$ singles out pairs with asymmetry energy $\Delta = \sqrt{\hbar^2\omega^2 - \eta^2}$, resulting in a fixed value for the non-diagonal part of the dipole moment (3.43).

Generation of Rotary Echoes

The generation of a rotary echo is schematically illustrated in Fig. 3.3. At low temperatures ($k_\mathrm{B}T < E$) the majority of the resonant tunneling systems are in the ground state. An r.f. field with frequency $\omega/2\pi$ and amplitude F induces dipolar transitions between the two levels. For $\omega = E/\hbar$ this leads to a periodic change of the occupation numbers and to a harmonic oscillation of the macroscopic electrical polarization with the Rabi frequency $\Omega_\mathrm{R} = (1/\hbar)P_\mathrm{eff} \cdot F$. In the present case, the effective dipole moment P_eff is given by the non-diagonal part (3.43); thus we have for a field parallel to the x axis,

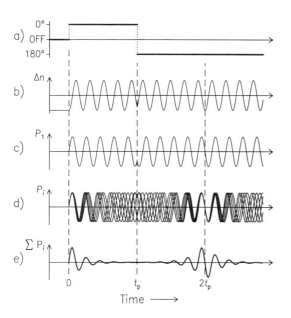

Fig. 3.3. Formation of a dielectric rotary echo as a function of time. (**a**) Phase of the applied r.f. field. (**b**) Time evolution of the occupation number difference of even and odd states of a coupled pair. (**c**) Time evolution of the corresponding polarization. (**d**) Superposition of the polarization of five tunneling systems with slightly different Rabi frequencies. (**e**) Time evolution of the resulting macroscopic polarization of tunneling systems with a Lorentzian distribution of Rabi frequencies. By courtesy of Enss and Weis [42]

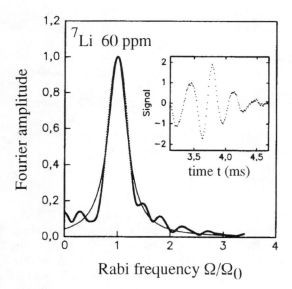

Fig. 3.4. Fourier transform of the amplitude of a rotary echo measured in KCl:^{7}Li$^+$. The solid line represents a numerical fit with a Lorentzian distribution of Ω/Ω_0, yielding $\Omega_0 = 2.93$ MHz. The inset shows the rotary echo used for the Fourier transformation. The echo has been obtained at $\omega/2\pi = 1125$ MHz and a field strength F of 201 V/m. By courtesy of Weis and Enss [42]

$$\Omega_R = \frac{1}{\hbar}\frac{\eta}{E}qdF. \tag{3.69}$$

(This expression accounts for coherent coupling of two impurities to the external field, with $P_x = p_x + p'_x$ and $|p_x| = \frac{1}{2}qd = |p'_x|$.)

For fixed tunnel energy η, the random asymmetry energy Δ leads to a continuous distribution for the energy splitting E, and hence to a band of Rabi frequencies contributing to the signal. Accordingly, the macroscopic polarization decays over time. However, when we invert the phase of the driving field at $t = t_p$, this is equivalent to a reversal of the time evolution of the polarization, and causes a revival of the Rabi oscillations at time $t = 2t_p$. This phenomenon is called a rotary echo.

Rotary Echoes of Pairs of Lithium Impurities

The inset of Fig. 3.4 shows a typical rotary echo pattern for a KCl crystal containing 60 ppm ^{7}Li. The time-dependent signal clearly displays oscillations with a frequency in the MHz range, and a phase coherence time much larger than the oscillation period. When we apply Fourier transformation of the time signal, we obtain the distribution of Rabi frequencies, which is sufficiently narrow to permit us to extract a meaningful average value.

The resulting Rabi frequency, as a function of field amplitude, is plotted in Fig. 3.5. Note that the data for the different resonance frequencies, 700 MHz and 1125 MHz, arise from distinct impurity pairs. The data confirm both the linear dependence on the field amplitude and the inverse variation with

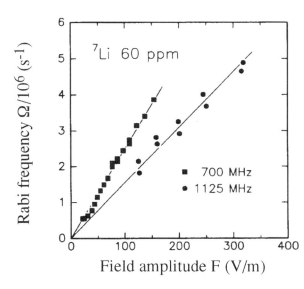

Fig. 3.5. Rabi frequencies Ω_R of ^{6}Li pairs in KCl as a function of the local electrical field strength F and for two different resonance frequencies. Data are taken at $T \simeq 10$ mK. By courtesy of Weis and Enss [42]

the resonance frequency, as is to be expected from (3.69). This means that the remaining factor $qd\eta/\hbar$ in (3.69) is constant; hence the pairs which are observed at different resonance frequencies have the same tunnel frequency $\eta/\hbar$.

Identical experiments have been performed on KCl-crystals containing 70 ppm ^{6}Li. Both the echo pattern and the Fourier spectrum look very similar to those of Fig. 3.4 [40,42], and the observed dependence of the Rabi frequency on resonance energy E and external field amplitude F agrees with (3.69).

When the sample is doped with equal parts of both isotopes, we are able to investigate pairs of one ^{6}Li$^+$ and one ^{7}Li$^+$ ion. For a statistical distribution of the isotopes, one should find twice as many mixed configurations as pure ^{7}Li or pure ^{6}Li pairs. According to (3.24), the tunneling energy of mixed pairs is expected to lie between the values for the pure isotopes, $^7\eta < {}^{6/7}\eta < {}^6\eta$. Figure 3.6 confirms that the linear field dependence holds true for the mixed pairs as well.

The straight lines in Figs. 3.5 and 3.6 agree with the parameters given in Table 3.2. The energy E is fixed by the external frequency ω; then the dipole strength, i.e., the slope in the figures, is determined by the tunnel energy η. The dipole moment p, the tunnel energy Δ_K, and the distance of next-nearest-neighbor sites are well-knwon; the only uncertainty arises from the effective dielectric constant, as discussed on p. 28. The polarizability of the host lattice leads to a screening effect; thus for impurities at sufficiently a large distance, one may use the proper dielectric constant $\epsilon = 4.49$. Yet for nearby impurities the screening is incomplete, and one has to use some

44   3. Tunneling of a Coupled Defect Pair

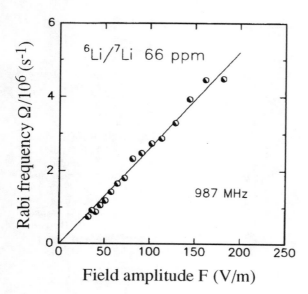

Fig. 3.6. Rabi frequencies Ω_R of mixed impurity pairs ^{6}Li/^{7}Li in KCl as a function of the local electrical field strength F at a resonance frequency $\omega/2\pi = 987$ MHz. Data are taken at $T \simeq 10$ mK. By courtesy of Weis and Enss [42]

effective value $1 \leq \epsilon_{\text{eff}} \leq \epsilon$ rather than the macroscopic quantity ϵ. Thus ϵ_{eff} is to be considered as a parameter; the data in Figs. 3.5 and 3.6, and similar ones for a sample doped with 70 ppm ^{6}Li, are well described with the value $\epsilon_{\text{eff}} = 2.25$.

In order to render more obvious the isotope effect, Fig. 3.7 displays the product of Rabi and resonance frequencies $\Omega_R\omega$. According to (3.24) and (3.69), one expects the slopes to be independent of $E = \hbar\omega$ and J, and to be entirely determined by the ratio $^6\Delta_K/^7\Delta_K$. The quantity $\Omega_R\omega$ thus provides a precise test for the dependence on the isotope mass. The straight lines in Fig. 3.7 are linear fits; in Table 3.3 we compare the ratios thus obtained with theoretical values for NN2 pairs. (We emphasize that the latter depend only on $^6\Delta_K$ and $^7\Delta_K$.)

Table 3.2. Parameters for the rotary echoes of Li pairs on next-nearest-neighbor sites NN2. The dipolar interaction J has been calculated from (3.7) with $R = 6.23$ Å; the value for the dielectric constant is discussed in the text.

ϵ_{eff}	p/D	J/K	$^6\Delta_K$/K	$^7\Delta_K$/K	$^6\eta$/mK	$^7\eta$/mK	$^{6/7}\eta$/mK
2.25	2.63	123	1.65	1.07	22.0	9.3	14.3

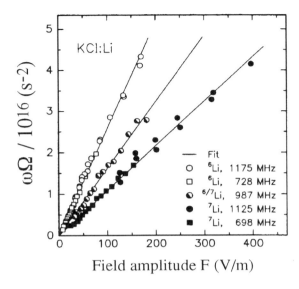

Fig. 3.7. The product $\Omega_R\omega$ as a function of the local electric field strength F for all three types of impurity pairs. By courtesy of Weis and Enss [42]

Figure 3.7 and Table 3.3 summarize the results of rotary echoes and permit us to draw several conclusions concerning the dependence on external field F, applied frequency ω, and isotope mass.

(a) Field Dependence. The theoretical description of rotary echoes requires us to calculate a non-linear response function. The energy spectrum and the selection rules shown in Fig. 3.2 and (3.46) permit us to use well-known results for rotary echoes of two-level systems [83–85]. The linear variation with the external field proves this approach to be correct.

(b) Frequency Dependence. For the samples containing either ^{6}Li or ^{7}Li, Fig. 3.7 shows data obtained at two different frequencies. The fact that the quantity $\Omega_R\omega$ is independent of the external frequency confirms the validity of our model and, in particular, the symmetry properties of the Hamiltonian.

(c) Isotope Effect. The variation with the impurity mass strongly supports the microscopic model for the tunneling motion developed in Sect. 3.2. The data of Table 3.3 exclude tunneling along face and space diagonals.

The surprisingly good quantitative agreement between theory and data proves that the observed rotary echoes arise from coherent tunneling oscillations of two strongly coupled Li$^+$ ions with next-nearest-neighbor configuration NN2. The oscillations in real time shown, in the insets of Fig. 3.4, correspond to coherent tunneling between the odd ($\sigma_x = 1$) and even ($\sigma_x = -1$) collective eigenstates of the two defects; this tunneling motion is driven by the time-dependent external field.

In view of the case of higher concentration to be treated in Chap. 5, we emphasize two conclusions to be drawn from the present results.

46   3. Tunneling of a Coupled Defect Pair

Table 3.3. Isotope effect on the tunnel energy. Theoretical values arise from $^6\Delta_{\rm K}/^7\Delta_{\rm K} = 1.54$, experimental data are given by the fits in Fig. 3.7

	$^6\eta\,/\,^{6/7}\eta$	$^{6/7}\eta\,/\,^7\eta$	$^6\eta\,/\,^7\eta$
Experimental	1.59	1.50	2.38
Theoretical	1.54	1.54	2.38

First, the static asymmetry arising e.g., from lattice imperfections is small; for the observed defect pairs Δ takes values of less than 30 mK. This indicates that the relaxation peak in the dynamic susceptibility found at higher lithium concentration does not arise from a static asymmetry energy but is of different origin, contrary to the situation encountered for tunneling systems in glasses.

Second, the theory presented here yields results similar to those obtained from the pair model of two coupled two-level systems, and thus shows the two-state approximation for interacting tunneling impurities to be essentially correct. This is of interest for the description of higher concentrated samples where one faces a complicated many-body problem and where the two-state approximation can hardly be avoided.

3.6 Discussion

Validity of Perturbation Theory

The energy spectrum of the ground state multiplet has been calculated to second order in terms of the perturbation $K + V$; the unperturbed Hamiltonian is given by the dipolar interaction W. Since $PVQ = 0$, the potential V contributes a first-order term only; on the other hand, when we note that $PKP = 0$, it is clear the first non-vanishing term arising from the tunneling part is of second order in K.

The eigenstates $|\sigma\rangle$ are confined to the subspace spanned by the ground states $|G\alpha\beta\gamma\rangle$. Formally this corresponds to a zeroth-order approximation; the first-order corrections would yield an admixture of excited states with energies E_1 and E_2 and amplitudes of the order of magnitude $\Delta_{\rm K}/J$. The resulting corrections to the susceptibility are immaterial for our purpose.

The perturbation analysis relies on the small parameters $\Delta_{\rm K}/J$ and Δ/J; there is no restriction on the ratio $\Delta_{\rm K}/\Delta$. The most relevant range, however, involves an asymmetry energy of the order of $\eta = \Delta_{\rm K}^2/J$ or smaller, i.e., $\Delta \ll \Delta_{\rm K}$. For lithium impurities on nearby sites one finds for the small parameter $\Delta_{\rm K}/J \approx 1/100$. Thus, perturbation theory should yield essentially exact results.

Degenerate perturbation theory in an eight-dimensional subspace may yield quite involved expressions for the energy spectrum and the eigenstates. Luckily, the present problem factorizes with respect to tunneling motion in the three crystal directions. As a consequence, the wave function may be written as a product of three factors, each depending on either x, y, or z [compare (3.33)]. When we attempt to go beyond the perturbation analysis, a more explicit use of the factorization property might prove advantageous [82].

Role of the Asymmetry Energy

Other impurities or additional interactions may remove the degeneracy of the eight off-center positions of a single defect. For a sufficiently distant source, this additional potential may be linearized in terms of the position operators $\boldsymbol{r}$ and $\boldsymbol{r}'$, resulting in a perturbation of the form (3.4). Owing to the particular form of the ground states NN2, the contributions from the two impurities add along the axis of the pair and cancel for the remaining directions.

Both energy spectrum and eigenstates are labeled by three quantum numbers $\boldsymbol{\sigma}$. For zero asymmetry, $\Delta = 0$, the variable σ_x denotes even ($\sigma_x = -1$) or odd ($\sigma_x = 1$) superpositions of the states localized at $x = \pm d/2 = x'$; similar statements apply for σ_y and σ_z. In the case of a finite bias Δ, the meaning of the label σ_x changes; it then denotes two different superpositions with amplitudes $\tilde{f}_\alpha^\sigma$, which in the limit of large asymmetry tend towards the partially localized states $|G\alpha_x; \sigma_y \sigma_z\rangle$.

Several features of the one-particle potential (3.4) are essential for the observation of rotary echoes:

(a) For a fixed value of the tunnel splitting η, the random configuration of the sources of the perturbation V leads to a continuous distribution of the asymmetry energy Δ, and hence of the energy splitting E. Therefore the resonance condition $\hbar\omega = E$ is satisfied for any frequency $\omega \geq \eta/\hbar$.

(b) Owing to its symmetry properties, (3.4) does not introduce additional transitions, but merely increases the relevant energy difference from η to E, as shown in Fig. 3.2.

(c) The asymmetry energy affects both the energy splitting and the dipolar strength of all three transitions in the same fashion, resulting in the particularly simple form for the response function (3.46).

A more general asymmetry potential with $\boldsymbol{K} \neq \boldsymbol{K}'$ in (3.4) would lead to a more complicated energy spectrum with various transition frequencies and dipole strengths. As a consequence, the resonance condition $\hbar\omega = E$ could be satisfied by quite different systems, resulting in a broad distribution of Rabi frequencies instead of the sharp spectrum observed (see Fig. 3.4).

These experiments [40–42] thus strongly support the present model for nearby pairs of impurities, and in particular confirm the symmetry properties of the asymmetry potential (3.4).

Isotope Effect

The availability of two stable isotopes, ^{6}Li and ^{7}Li, constitutes a most appealing aspect of lithium defects. Defect pairs may involve either two ions of the same mass or two different isotopes. According to the WKB formula $\Delta_K = \hbar\omega_0 \exp(-(d/\hbar)\sqrt{2mV_0})$, the tunneling amplitude depends exponentially on the defect mass, whereas the dipole moment is the same for both isotopes. Thus we have accounted for different tunneling parts T and T' only.

The diagonal part of $\tilde{H}$ involves η_0, (3.23), and merely causes an insignificant shift of the energy spectrum (3.30); the more interesting non-diagonal part results in the tunnel splitting which is determined by the quantity $\eta = \Delta_K \Delta'_K / J$. When we consider all combinations of $^6\Delta_K$ and $^7\Delta_K$, we obtain the isotope effect for both resonance energy (3.37) and off-diagonal dipolar transition amplitudes (3.43).

In Table 3.3 we have gathered the values for the ratio of the effective two-impurity tunnel energy η, as derived from the rotary echo data. The 'theoretical' values have been calculated from (3.24), with $^6\Delta_0$ and $^7\Delta_0$ as obtained from specific heat measurements and other techniques [41].

According to Table 3.5, the dipolar interaction J is at least one hundred times larger than the tunnel energy Δ_K. One would not be surprised to find that the tunnel states described in Chap. 2 are significantly distorted by the interaction. (The barrier height between different off-center positions is approximately 40 K, which is significantly smaller than the coupling energy J.) Since the tunnel splitting Δ_K is higly dependent on the shape of the potential, one would expect the ratio $^6\Delta_K/^7\Delta_K$ to vary with increasing interaction, i.e., for pairs to get closer and closer. Yet Table 3.3 states this ratio to be same for non-interacting defects and for pairs on next-nearest-neighbor sites at a distance $R = 6.23$ Å.

Angle Dependence of the Rabi Frequency

Rotary echoes permit us to observe the non-linear response to an external, field $\boldsymbol{F}\cos(\omega t)$. Up to now we have assumed that the field vector $\boldsymbol{F}$ is parallel to one of the crystal axes. For a NN2 pair along the x axis, the susceptibility tensor (3.36) shows a single finite entry α_{xx}, as calculated in (3.46); for a field $\boldsymbol{F} = \boldsymbol{e}_x F$ only systems of this type contribute to the rotary echo, with the Rabi frequency (3.69).

A different situation is encountered when applying an external field pointing in an arbitrary direction,

$$\boldsymbol{F} = (F_x, F_y, F_z). \tag{3.70}$$

Owing to the random configuration of the impurities, a third of the resonant pairs (i.e., those satisfying $\hbar\omega = E$) will be oriented along each of the three axes and experience the respective field amplitudes F_x, F_y, and F_z. According to (3.69), these give rise to three different Rabi frequencies. (Putting $F_y = 0 = F_z$ one easlily recovers the particular case studied in this chapter.)

The experiments discussed above were carried out on single crystals; changing the orientation of the electric field with respect to the crystal axes should permit us to observe a variation in the Rabi frequency, and a superposition of echoes with different frequencies for the general case.

4. A Pair of Two-Level Systems

The problem of interacting tunneling defects may be significantly simplified by reducing the number of degenerate positions for each impurity; one widely used model is based on an effective two-state system mimicking the actual eight off-center positions of a single defect ion. Here the introduction of this two-state approximation serves two purposes.

First, for weakly doped defect crystals static quantities, such as the specific heat, are mainly affected by defect pairs with moderate coupling energy for which the perturbative approach of the previous section ceases to be valid. The two-state approximation allows us to find an exact solution for a defect pair, thus providing a proper description of the specific heat of samples with not more than about 100 ppm lithium defects, or about 600 ppm cyanide impurities.

Second, at higher concentrations, the Hamiltonian for N impurities cannot be split into $N/2$ pairs that are independent of each other, but one has rather to face a N-particle problem. Yet the treatment of the latter in Chap. 5 relies heavily on the two-state approximation for each impurity. Comparison with the results from perturbation theory for two interacting eight-state systems permits us to make an estimate of the validity of the two-state approximation.

4.1 Two-State Approximation

The essential feature consists in replacing the eight minima of (2.2) by a double-well potential with localized states $|L\rangle$ and $|R\rangle$. (A simple example is provided by $V(x, y_0, z_0)$ as a function of x, with y_0 and z_0 remaining constant.) When we identify the tunneling amplitude $-(\Delta_0/2) = \langle L|V|R\rangle$ with the largest term in (2.3),

$$\Delta_0 = \Delta_K, \qquad (4.1)$$

and define Pauli matrices

$$\sigma_x = |L\rangle\langle R| + |R\rangle\langle L| \quad \text{and} \quad \sigma_z = |L\rangle\langle L| - |R\rangle\langle R|, \qquad (4.2)$$

we obtain the effective two-state Hamiltonian

$$T = -\tfrac{1}{2}\Delta_0 \sigma_x, \tag{4.3}$$

whose eigenstates $2^{-1/2}(|L\rangle \pm |R\rangle)$ are separated by the tunnel energy Δ_0. The corresponding internal energy

$$U = -\tfrac{1}{2}\Delta_0 \tanh(\Delta_0/2k_\mathrm{B}T) \tag{4.4}$$

differs from (2.15) by a factor of 3. In order to adjust the results from the two-state approximation to those for the actual eight-state system, we are thus led to multiply the internal energy by a factor of 3.

With the constant vector $\boldsymbol{d}$, the two values of the position operator

$$\boldsymbol{r} = \tfrac{1}{2}\boldsymbol{d}\sigma_z \tag{4.5}$$

correspond to the localized quantum states $|L\rangle$ and $|R\rangle$ labeled by $\sigma_z = \pm 1$. Moreover, we define the dipole moment

$$\boldsymbol{p} = \frac{1}{2} q \boldsymbol{d}\sigma_z, \tag{4.6}$$

whose absolute value reads with $|\boldsymbol{d}| = d$

$$|\boldsymbol{p}| = \frac{1}{2} q d = p. \tag{4.7}$$

The dynamic response function with respect to an electric field parallel to $\boldsymbol{d}$ is given by

$$\alpha(t-t') = \frac{2}{\hbar}\frac{p^2}{\epsilon_0}\tanh(\Delta_0/2k_\mathrm{B}T)\sin[\Delta_0(t-t')/\hbar]\Theta(t-t'). \tag{4.8}$$

Calculation of the response function of N impurities in a sample of volume V requires us to average over the dipole orientation with respect to the applied field and to sum over all defects,

$$\chi_\mathrm{iso}(t-t') = \frac{1}{3V}\sum_i \alpha_i(t-t'). \tag{4.9}$$

(The factor $\tfrac{1}{3}$ arises from the orientational average, and the label indicates non-interacting two-level systems, isolated from each other.)

Taking the Fourier transform and using $n = N/V$ finally yields the corresponding dynamic susceptibility

$$\chi_\mathrm{iso}(\omega) = \frac{2}{3}\frac{np^2}{\epsilon_0}\frac{\Delta_0}{\Delta_0^2 - \hbar^2\omega^2}\tanh(\Delta_0/2k_\mathrm{B}T), \tag{4.10}$$

Comparison with the result for the actual eight-state system, (2.24), reveals the formal identity of both expressions. This coincidence makes us confident in treating interacting tunneling defects in two-state approximation.

For later use, we write down explicitly the susceptibility for the case of zero frequency,

$$\chi_\mathrm{iso}(\omega = 0) = \frac{2}{3}\frac{np^2}{\epsilon_0 \Delta_0}\tanh(\Delta_0/2k_\mathrm{B}T). \tag{4.11}$$

When dealing with interacting impurities in the sequel, we will constantly use (4.11) as a reference for the low-density limit. Although its derivation involves some arbitrary numerical constants, (4.10) and (4.11) are identical to the exact expressions (2.24) and (2.27).

4.2 The Pair Model

In terms of the number density $n = N/V$, the average distance of adjacent impurities is roughly given by $n^{-1/3}$. Their dipolar interaction decreases with the third power of distance; thus at sufficiently low density most impurities experience a negligible dipolar field. A random distribution of the defects on the host lattice, however, implies the occurrence of a few clusters of nearby defects with large coupling energy, as is shown schematically in the middle part of Fig. 1.2 on page 5. When we truncate a virial expansion at its first nontrivial contribution, we obtain $N/2$ pairs of impurities; higher-order terms involve corresponding powers of n and hence are negligible at low densities.

This two-level model for a pair of tunneling impurities was invented by Baur and Salzman [48] when investigating potassium chloride doped with Li, CN, or OH ions. In a series of papers, Klein refined the model and calculated the energy spectrum and various related quantities [49]. Description of dynamic experiments however requires knowledge of time correlations. Kranjc solved the dynamic equation of a pair of two-level defects by considering the tunneling motion as a small perturbation [50]. The solution presented here draws heavily on work by Terzidis, who has obtained the exact dynamics and the statistical operator for an impurity pair [51,52].

The replacement of the coupled impurities of the last chapter by pairs of two-level systems involves two further steps, besides the tunnel energy (4.1) and the dipole moment (4.6); we still have to specify the total dipole moment

$$\boldsymbol{P} = \tfrac{1}{2}q(\boldsymbol{d}\sigma_z + \boldsymbol{d}'\sigma'_z) \tag{4.12}$$

and the interaction energy of a pair of two-level systems with distance $\boldsymbol{R}$. The latter is obtained by inserting the respective dipole moments (4.6) in Eq. (3.5). When using (3.7) and defining unit vectors through $\boldsymbol{d} \equiv d\hat{\boldsymbol{d}}$ etc, we find the interaction term

$$W = \tfrac{1}{2}J\left(\tfrac{1}{2}\hat{\boldsymbol{d}}\cdot\hat{\boldsymbol{d}}' - \tfrac{3}{2}(\hat{\boldsymbol{d}}\cdot\hat{\boldsymbol{R}})(\hat{\boldsymbol{d}}'\cdot\hat{\boldsymbol{R}})\right)\sigma_z\sigma'_z. \tag{4.13}$$

If one deals with physical two-state dipole moments, the relative orientation of the vectors $\boldsymbol{d}$, $\boldsymbol{d}'$, and $\boldsymbol{R}$ determine both dipole moment $\boldsymbol{P}$ and interaction energy J; note that the expression in brackets may take both signs.

Any of the three configurations discussed in the previous chapter could be accounted for by an appropriate choice of parameters. Yet since our present purpose is to perform the configurational average, we merely retain the variation with the distance R, and discard the dependence on the relative orientation. In order to meet the most relevant results for configuration NN2, we put

$$\boldsymbol{P} = \tfrac{1}{2}qd(\sigma_z + \sigma_z') = p(\sigma_z + \sigma_z'), \tag{4.14}$$

and we replace the term in brackets in (4.13) by ± 1. With

$$J = \frac{1}{4\pi\epsilon_0\epsilon}\frac{q^2d^2}{R^3}, \tag{4.15}$$

as defined in (3.7), we have

$$W = -\tfrac{1}{2}J\sigma_z\sigma_z'. \tag{4.16}$$

In order to account for the ground states found in the previous chapter, the sign in (4.16) has been chosen so that with $J > 0$ it favors parallel alignment of the dipole moments ('parallel' here means $\sigma_z\sigma_z' = 1$). The sign of J is not an irrelevant matter; the response function for strongly interacting pairs turns out to differ significantly for the two signs. (In order not to encumber the notation, we consider mainly the physical case $J > 0$ in the sequel, and indicate where the sign of J would modify the results.)

Assuming equal tunneling amplitudes $\Delta_0 = \Delta_0'$ and discarding an asymmetry term $\tfrac{1}{2}\Delta(\sigma_z + \sigma_z')$, we obtain the Hamiltonian for an interacting pair

$$H = -\tfrac{1}{2}\Delta_0\sigma_x - \tfrac{1}{2}\Delta_0\sigma_x' - \tfrac{1}{2}J\sigma_z\sigma_z'. \tag{4.17}$$

(The more general case $\Delta_0 \neq \Delta_0'$ is treated in [51]; for the effect of a finite asymmetry energy Δ, see [53].) A natural basis for the four-dimensional space is provided by the product states

$$|LL\rangle, |RR\rangle, |LR\rangle, |RL\rangle; \tag{4.18}$$

both dipole moment and interaction W are diagonal in this basis.

From the four states (4.5) one may construct 16 quantum-mechanical operators; using (4.2) we choose the particular set

$$\begin{array}{llllll}
(1) & \sigma_y, & \sigma_z, & \sigma_x\sigma_y', & \sigma_x\sigma_z', & \\
(2) & \sigma_y', & \sigma_z', & \sigma_y\sigma_x', & \sigma_z\sigma_x', & \\
(3) & 1, & \sigma_x\sigma_x', & & & \\
(4) & \sigma_x, & \sigma_x', & \sigma_y\sigma_y', & \sigma_y\sigma_z', & \sigma_z\sigma_y', & \sigma_z\sigma_z'.
\end{array} \tag{4.19}$$

These operators form a closed algebra: The product of any two elements is again a member of the set.

4.3 The Statistical Operator

The Hamiltonian is invariant under the parity transformation Π: $\boldsymbol{r} \to -\boldsymbol{r}$, $\boldsymbol{r}' \to -\boldsymbol{r}'$, which, in terms of quantum states, reads $\Pi = \sigma_x\sigma_x'$. One easily verifies $[\Pi, H] = 0$; accordingly, there are at least two integrals of motion, H and Π. In the basis (4.18), the Hamilton matrix reads

$$H = \begin{pmatrix} -J & 0 & -\Delta_0 & -\Delta_0 \\ 0 & -J & -\Delta_0 & -\Delta_0 \\ -\Delta_0 & -\Delta_0 & J & 0 \\ -\Delta_0 & -\Delta_0 & 0 & J \end{pmatrix}; \quad (4.20)$$

its eigenvalues are given by the roots of the the secular equation

$$0 = \det(E - H) = \left(E^2 - \tfrac{1}{4}J^2\right)\left(E^2 - \Delta_0^2 - \tfrac{1}{4}J^2\right). \quad (4.21)$$

In Fig. 4.1, we plot the four levels resulting from (4.21) as functions of the ratio J/Δ_0. For later use we define the energy difference of the ground state to the first excited level η_-, and to the second excited level η_+, which read explicitly

$$\eta_\pm = \sqrt{\Delta_0^2 + \tfrac{1}{4}J^2} \pm \tfrac{1}{2}J. \quad (4.22)$$

(With $\eta_+ \eta_- = \Delta_0^2$, it is clear that a change of sign for J amounts to exchanging η_+ and η_- [52].)

For zero interaction, the four states (4.18) form three levels with energies $-\Delta_0$, 0, Δ_0 and degeneracies 1, 2, 1. In the strong coupling limit $J \gg \Delta_0$, the spectrum consists of two doublets separated by an energy $\eta_+ \approx J$; each doublet displays a splitting $\eta_- \approx \Delta_0^2/J$.

Calculation of thermal averages requires knowledge of the statistical operator

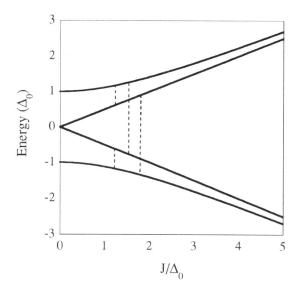

Fig. 4.1. Energy levels of the pair model as calculated from (4.21). The dashed lines indicate allowed electric dipole transitions with resonance frequencies $\eta_-/\hbar$ and $\eta_+/\hbar$ as given by (4.22)

56   4. A Pair of Two-Level Systems

$$\rho = e^{-\beta H}; \tag{4.23}$$

its diagonalization is carried out in two steps. First we transform the basis to eigenstates of the interaction term (4.16),

$$2^{-1/2}\left(|LL\rangle \mp |RR\rangle\right), \quad 2^{-1/2}\left(|LR\rangle \pm |RL\rangle\right) \tag{4.24}$$

which is achieved by the orthogonal transformation

$$U = \frac{1}{\sqrt{2}} \begin{pmatrix} -1 & 1 & 0 & 0 \\ 1 & 1 & 0 & 0 \\ 0 & 0 & 1 & 1 \\ 0 & 0 & 1 & -1 \end{pmatrix}. \tag{4.25}$$

The transformed Hamiltonian matrix $U^\dagger H U \equiv \tilde{H}$,

$$\tilde{H} = \begin{pmatrix} -J & 0 & 0 & 0 \\ 0 & -J & -2\Delta_0 & 0 \\ 0 & -2\Delta_0 & J & 0 \\ 0 & 0 & 0 & J \end{pmatrix}, \tag{4.26}$$

allows us to write the statistical operator as

$$\rho = U e^{-\beta \tilde{H}} U^\dagger; \tag{4.27}$$

after splitting off a factor arising from the diagonal part of $\tilde{H}$, we are left with a 2×2 matrix

$$\tilde{h} = \begin{pmatrix} -J & -2\Delta_0 \\ 2\Delta_0 & J \end{pmatrix} \tag{4.28}$$

in the exponential. By inserting $Q = \sqrt{4\Delta_0^2 + J^2}$ and

$$\exp(-\beta \tilde{h}) = \cosh\left(\tfrac{1}{2}\beta Q\right) - \sinh\left(\tfrac{1}{2}\beta Q\right) Q^{-1} \tilde{h}, \tag{4.29}$$

performing back-transformation by means of (4.25), and expansion in terms of (4.19), Terzidis calculated the statistical operator [51–53]

$$\rho = 1 + w_1\, \sigma_x \sigma'_x + w_2\, \sigma_x + w_2\, \sigma'_x + w_3\, \sigma_y \sigma'_y + w_4\, \sigma_z \sigma'_z. \tag{4.30}$$

The coefficients

$$\begin{aligned}
w_1 &= \tanh(\beta \eta_-/2)\tanh(\beta \eta_+/2), \\
w_2 &= -\sqrt{R_+ R_-}\left[\tanh(\beta \eta_-/2) + \tanh(\beta \eta_+/2)\right], \\
w_3 &= R_+ \tanh(\beta \eta_+/2) - R_- \tanh(\beta \eta_-/2), \\
w_4 &= R_- \tanh(\beta \eta_+/2) - R_+ \tanh(\beta \eta_-/2),
\end{aligned} \tag{4.31}$$

are determined by the energies (4.22) and the amplitudes

$$R_\pm = \frac{\eta_\mp}{\eta_+ + \eta_-} = \frac{\Delta_0^2}{\Delta_0^2 + \eta_\pm^2}. \tag{4.32}$$

In the latter equation we have used $\eta_- \eta_+ = \Delta_0^2$. Note $\operatorname{tr}\rho = 4$.

According to the remark below Eq. (4.22) and considering the expression for $R_\pm$, the dependence of the coefficients $w_1, ..., w_4$ on the sign of J is easily determined. Putting $\eta_\pm \to \eta_\mp$ and with (4.32) $R_\pm \to R_\mp$, we find w_1 and w_2 to be invariant, whereas w_3 and w_4 take different signs for $J > 0$ and $J < 0$.

For weak coupling $|J| \ll \Delta_0$, the two terms of w_4 are of comparable absolute value; because of the different sign, they cancel almost completely. In the case of strong coupling we find $w_4 \approx \pm 1$; since w_4 always carries the same sign as J, we find

$$w_4 = \text{sign}(J)|w_4|, \tag{4.33}$$

which will be useful when discussing the dynamic susceptibility.

4.4 Density of States

In order to find the average over disorder, i.e., over the random defect configuration, we need the distribution of the energies $\eta_\pm$

$$\rho_{\text{PM}}(E) = \frac{1}{2N} \sum_i \sum_\pm \delta\left(E - \eta_\pm^{(i)}\right). \tag{4.34}$$

[The factor $(1/2N)$ is included for normalization.] Since a given impurity interacts with many others, the definition of a pair is not unambiguous; yet in the dilute case, the strong distance dependence of the interaction energy provides a sound criterion. The random impurity configuration on the host

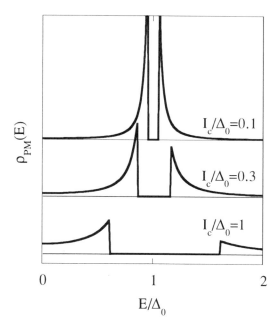

Fig. 4.2. Density of states $\rho_{\text{PM}}(E)$ according to (4.37) for various values of the parameter I_c, corresponding to different impurity densities

lattice leads to a distribution function for the coupling energy J which is given by (C26),

$$\tilde{Q}(J) = I_c/J^2 \quad \text{for } I_c \leq J \leq J_2, \tag{4.35}$$

with the cut-off value

$$I_c = \frac{8\pi}{3} \frac{np^2}{4\pi\epsilon\epsilon_0}. \tag{4.36}$$

When we note $|dJ/d\eta_\pm| = (\eta_\pm^2 + \Delta_0^2)/2\eta_\pm^2$ and $n = N/V$, Klein's result [49] for the density of states reads in our notation

$$\rho_{\text{PM}}(E) = \frac{I_c}{2} \frac{E^2 + \Delta_0^2}{(E^2 - \Delta_0^2)^2} \quad \text{for } \begin{cases} \Delta_0^2/J_2 \leq E \leq \sqrt{\Delta_0^2 + \tfrac{1}{4}I_c^2} - \tfrac{1}{2}I_c \\ \sqrt{\Delta_0^2 + \tfrac{1}{4}I_c^2} + \tfrac{1}{2}I_c \leq E \leq J_2. \end{cases} \tag{4.37}$$

In Fig. 4.2 we plot the function $\rho_{\text{PM}}(E)$ for three values of I_c/Δ_0. For ^{7}Li, the case $I_c/\Delta_0 = 0.1$ corresponds to an impurity concentration of about 70 ppm; when we increase the defect density, the result is a broadening of the gap about $E = \Delta_0$ and an enhancement of the density of states at low energies.

For later use we note the coupled density of states

$$D_{\text{PM}}(E) = \frac{1}{N} \sum_i \sum_\pm R_\pm \delta\left(E - \eta_\pm^{(i)}\right), \tag{4.38}$$

where $R_\pm$ is given by the amplitudes defined in Eq. (4.32); one finds

$$D_{\text{PM}}(E) = I_c \frac{\Delta_0^2}{(E^2 - \Delta_0^2)^2} \tag{4.39}$$

with the same restrictions on E as in (4.37).

Both $\rho_{\text{PM}}(E)$ and $D_{\text{PM}}(E)$ are normalized to unity, $\int dE \rho_{\text{PM}}(E) = 1$ and $\int dE D_{\text{PM}}(E) = 1$; we have assumed $\Delta_0 \ll J_2$ and $c \ll 1$, and neglected small quantities accordingly.

4.5 Specific Heat

From the energy eigenvalues (4.22), one easily calculates the partition function for a pair with interaction energy J,

$$Z = \cosh(\beta\eta_+/2)\cosh(\beta\eta_-/2), \tag{4.40}$$

and the internal energy $U = -\partial_\beta \log(Z)$,

$$U = -\tfrac{1}{2}\eta_+ \tanh(\beta\eta_+/2) - \tfrac{1}{2}\eta_- \tanh(\beta\eta_-/2). \tag{4.41}$$

When attempting to calculate static quantities such as the specific heat, one has to average over all impurities; in the pair model this amounts to considering pairs with different distances and to performing both the thermal average and that over distance,

$$C_V = 3\frac{1}{V}\sum_{\text{pairs}} \frac{dU}{dT}, \tag{4.42}$$

[For the factor 3, see the discussion after (4.4).]

When we use the variable $x = \eta_\pm/\Delta_0$ and the corresponding weight function $\tilde{\rho}_{\text{PM}}(x) = (I_c/2\Delta_0)(1+x^2)/(1-x^2)^2$, we obtain for the specific heat the integral

$$C_V = 3k_\text{B} n \left[\int_{x_0}^{x_-} dx + \int_{x_+}^{x_0^{-1}} dx \right] \tilde{\rho}_{\text{PM}}(x) \frac{(\beta\Delta_0 x/2)^2}{\cosh(\beta\Delta_0 x/2)^2}, \tag{4.43}$$

whose upper and lower bounds are given by

$$x_0 = \Delta_0/J_2, \qquad x_\pm = \sqrt{1 + \tfrac{1}{4}(I_c/\Delta_0)^2} \pm \tfrac{1}{2}(I_c/\Delta_0). \tag{4.44}$$

Polar molecules such as CN and off-center impurities such as Li are equally well described by the eight-state model of Chap. 2; accordingly, the two-state approximation and the pair model introduced in the present chapter apply to both defect systems.

In Fig. 4.3 we show specific heat data for KCl with 27 and 621 ppm of cyanide impurities [54]. The dashed lines give the temperature dependence expected for non-interacting impurities according to (2.16), whereas the full lines account for the pair model [54]. The data on the sample with 27 ppm

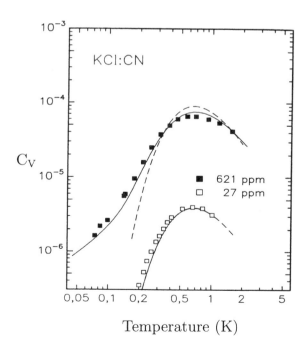

Fig. 4.3. Specific heat for 27 and 621 ppm cyanide impurities in KCl, in units of (J/gK). The Debye contribution of the host crystal has been subtracted. The data are from Peressini et al. [54]. Drawing by courtesy of Weis [41]

4. A Pair of Two-Level Systems

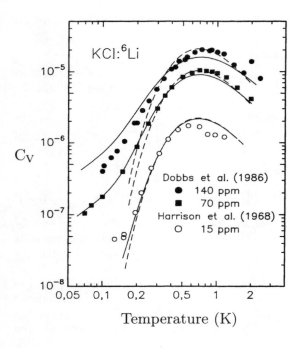

Fig. 4.4. Specific heat for 15, 70, and 140 ppm ^{6}Li impurities in KCl, in units of (J/gK). The Debye contribution of the host crystal has been subtracted. Data from [45] and [57]. Drawing by courtesy of Weis [41]

CN are well described by the Schottky curve, (2.16), which in fact differs little from that obtained with the pair model. As compared to lithium ions, the cyanide impurities interact only weakly; thus at a concentration of 621 ppm, the average coupling energy is still smaller than the tunnel energy $\Delta_0/k_B = 1.6$ K. Figure 4.3 displays a clear Schottky peak for the 621 ppm data; the excess specific heat below 200 mK is due to the smaller energy η_- of sufficiently close pairs. Similar data have been reported in [56].

The effect of interaction on the specific heat of lithium impurities is displayed in Fig. 4.4. Again, dashed and solid lines give the Schottky curve (2.16) and the behavior obtained from the pair model, (4.43). Whereas the 15 ppm data are reasonably well describe by a single Schottky peak, the samples with higher impurity concentration exhibit clear deviations. The 70 ppm data are in perfect accord with the pair model; note the excess specific heat at low temperatures arising from the small-energy excitations of nearby defect pairs. The less good agreement for the sample with 140 ppm indicates the onset of interaction with more than one neighbor; it will be discussed in the next chapter. (This discussion applies equally well to data reported in [54,57,58].)

4.6 Dynamic Susceptibility

Time evolution of any operator is governed by the Heisenberg equation

$$\hbar \dot{A}(t) = \mathrm{i}[H, A(t)] \equiv \mathrm{i}\hbar \mathcal{L} A(t), \tag{4.45}$$

where the Liouvillian $\mathcal{L}$ acts as a linear operator on the space spanned by (4.19). The sixteen basis elements $\{A_i\}$ are arranged in such a way that each of the four groups forms an invariant subspace. When we note $\frac{1}{4}\mathrm{tr}(A_i A_j) = \delta_{ij}$ and define a matrix $\mathcal{L}_{ij} = \frac{1}{4}\mathrm{tr}(A_i \mathcal{L} A_j)$, we find the latter to be block-diagonal

$$\mathcal{L} = \begin{pmatrix} \mathcal{L}_1 & 0 & 0 & 0 \\ 0 & \mathcal{L}_2 & 0 & 0 \\ 0 & 0 & \mathcal{L}_3 & 0 \\ 0 & 0 & 0 & \mathcal{L}_4 \end{pmatrix}. \tag{4.46}$$

[The labels $i = 1, ..., 4$ correspond to the four subsets in (4.19).] $\mathcal{L}_1$ acts on the space spanned by $\sigma_y, \sigma_z, \sigma_x \sigma'_y, \sigma_x \sigma'_z$. A simple calculation yields

$$\mathcal{L}_1 = \frac{1}{\hbar} \begin{pmatrix} 0 & -\mathrm{i}\Delta_0 & 0 & \mathrm{i}J \\ \mathrm{i}\Delta_0 & 0 & 0 & 0 \\ 0 & 0 & 0 & -\mathrm{i}\Delta_0 \\ -\mathrm{i}J & 0 & \mathrm{i}\Delta_0 & 0 \end{pmatrix}. \tag{4.47}$$

$\mathcal{L}_2$ acts on the second subset of (4.19) and is identical to $\mathcal{L}_1$; $\mathcal{L}_3$ is equal to zero; the remaining part $\mathcal{L}_4$ is irrelevant for our purpose. The subspace belonging to eigenvalue zero is spanned by the operators

$$1, \quad H, \quad \sigma_x \sigma'_x, \quad H\sigma_x \sigma'_x, \tag{4.48}$$

which is obvious from the above discussion of the integrals of motion $\Pi = \sigma_x \sigma'_x$ and H. (A quantum mechanical operator A corresponds to an integral of motion if $\mathcal{L}A = 0$.)

The equation of motion (4.45) is solved by using Laplace transformation, which yields the resolvent representation

$$A(z) = \frac{-1}{z + \mathcal{L}} A. \tag{4.49}$$

[The notation $A(z)$ is quite formal; note that $A(z)$ is not proportional to the operator A, e. g., $\mathrm{tr}(AB) = 0$ does not imply $\mathrm{tr}(A(z)B) = 0$.] A basic dynamic quantity is given by the symmetrized two-time correlation function

$$G(t - t') \equiv \frac{1}{2} \langle \sigma_z(t) \sigma_z(t') + \sigma_z(t') \sigma_z(t) \rangle, \tag{4.50}$$

whose Laplace transform with respect to $t - t'$ reads as

$$G(z) = \frac{1}{2} \langle \sigma_z(z) \sigma_z + \sigma_z \sigma_z(z) \rangle. \tag{4.51}$$

After inversion of the matrix $z + \mathcal{L}_1$ we obtain

$$\sigma_z(z) = \frac{-1}{|z+\mathcal{L}_1|}\Big(\Delta_0(z^2 - \hbar^{-2}\Delta_0^2)\sigma_y - iz(z^2 - \hbar^{-2}\Delta_0^2 - \hbar^{-2}J^2)\sigma_z$$
$$+ \hbar^{-3}J\Delta_0^2\sigma_x\sigma_y' + iz\hbar^{-2}J\Delta_0\sigma_x\sigma_z'\Big); \tag{4.52}$$

using the cyclic invariance of the trace finally yields
$$G(z) = \frac{-z^3 - z(\Delta_0^2 - J^2)/\hbar^3}{z^4 - z^2(2\Delta_0^2 + J^2)/\hbar^2 + \Delta_0^4/\hbar^4}. \tag{4.53}$$

The determinant $|z+\mathcal{L}_1| = z^4 - z^2(2\Delta_0^2 + J^2)/\hbar^2 + \Delta_0^4/\hbar^4$ has four roots $\pm\eta_+$ und $\pm\eta_-$, which are given by (4.22). Singling out the pole contributions und using the amplitudes defined in (4.32), we get
$$G(z) = \sum_{\pm} \frac{\eta_\mp}{\eta_\pm + \eta_\mp} \frac{-z}{z^2 - \eta_\pm^2/\hbar^2} \equiv \sum_{\pm} R_\pm \frac{-z}{z^2 - \eta_\pm^2/\hbar^2}. \tag{4.54}$$

Laplace back-transformation yields oscillations with frequencies $\eta_\pm/\hbar$,
$$G(t-t') = \sum_{\pm} R_\pm \cos[\eta_\pm(t-t')/\hbar]. \tag{4.55}$$

When we note $\eta_\pm = \Delta_0 \pm \frac{1}{2}J$ for $J \ll \Delta_0$ and $\eta_+ = J$, $\eta_- = \Delta_0^2/J$ for $\Delta_0 \ll J$, we obtain simple expressions for $G(t-t')$ for these limits: in the weak-coupling case $J \ll \Delta_0$, both amplitudes are close to $\frac{1}{2}$, and the tunneling frequency is roughly given by $\Delta_0/\hbar$. For strong coupling $\Delta_0 \ll J$, the amplitude of the large frequency $J/\hbar$ is negligible, whereas that of the small frequency, $\Delta_0^2/J\hbar$, is close to unity.

Acoustic and microwave spectroscopy of tunnel excitations involves linear coupling to both impurities; assuming $\boldsymbol{d} = \boldsymbol{d}'$, we thus have to consider correlations of the operator
$$f = \sigma_z + \sigma_z'. \tag{4.56}$$

The corresponding two-time correlation function
$$S(t-t') = \tfrac{1}{2}\langle f(t)f(t') + f(t')f(t)\rangle \tag{4.57}$$

involves cross-correlations of the two pseudo-spins, besides the diagonal terms. When we take $\sigma_z = \sigma_z^1$, $\sigma_z' = \sigma_z^2$, and
$$G_{ij} = \tfrac{1}{2}\langle \sigma_z^i(t)\sigma_z^j(t') + \sigma_z^j(t')\sigma_z^i(t)\rangle, \tag{4.58}$$

we have $S(\tau) = \sum_{ij} G_{ij}(\tau)$. Both G_{11} and G_{22} are identical to (4.55). When we proceed as above, we find for the remaining terms
$$G_{12}(\tau) = G_{21}(\tau) = w_4 G_{11}(\tau), \tag{4.59}$$

thus with (4.55-4.58) we obtain
$$S(\tau) = 2(1+w_4)G(\tau). \tag{4.60}$$

The factor of 2 accounts for the fact that there are two impurities, whereas the additional term proportional to w_4 arises from interference of the waves scattered from the two particles.

4.6 Dynamic Susceptibility

Finally we turn to the dynamic response function

$$\alpha(t-t') = (\mathrm{i}/\hbar\epsilon_0)\langle[\boldsymbol{P}(t),\boldsymbol{P}(t')]\rangle\Theta(t-t'), \tag{4.61}$$

where the total dipole moment $\boldsymbol{P}$ is given by (4.14). Proceeding as for $S(t-t')$, or using a fluctuation-dissipation theorem for the spectra, $\alpha''(\omega) = (2p^2/\hbar\epsilon_0)S''(\omega)\tanh(\beta\hbar\omega)$, one finds

$$\alpha(\tau) = \frac{4}{\hbar}\frac{p^2}{\epsilon_0}(1+w_4)\sum_{\pm} R_{\pm}\tanh(\beta\eta_{\pm}/2)\sin(\eta_{\pm}\tau/\hbar)\Theta(\tau). \tag{4.62}$$

Now we discuss the effect of the sign of J, arising from the factor $(1+w_4)$, according to (4.31–4.33). In the weak-coupling limit $J \to 0$, the quantity w_4 vanishes linearly with J, and $\eta_{\pm}$ both tend towards Δ_0; thus for non-interacting impurities, (4.62) is just twice the single-particle susceptibility (4.8), as it should be.

Considering the case of low temperatures and a large absolute value for the coupling constant J, we find w_4 to be roughly equal to the sign, $w_4 \approx \mathrm{sign}(J)$, and in particular for negative J

$$1 + w_4 \approx 0 \qquad \text{for } k_\mathrm{B}T, \Delta_0 \ll -J. \tag{4.63}$$

Thus for strong negative coupling $J < 0$, the linear-response amplitude vanishes, and the impurity pair is essentially invisible at low temperature $k_\mathrm{B}T, \Delta_0 \ll -J$. (In general, w_4 may assume any value between -1 and 1.)

Finally we turn to the case $J \gg \Delta_0$, which results in $w_4 \approx 1$ and in an enhancement of the dielectric response by a factor of $(1+w_4) \approx 2$. According to the discussion at the beginning of this chapter, $J \gg \Delta_0$ corresponds to the strongly coupled pairs investigated in Chap. 3. In order to permit comparison with the results obtained there, we evaluate $\alpha(\tau)$ for $\Delta_0 \ll J$ and low temperatures $k_\mathrm{B}T \ll J \approx \eta_+$; neglecting small corrections to both amplitudes and frequencies, we find with $w_4 \approx 1$

$$\alpha(\tau) = (8/\hbar)(p^2/\epsilon_0)\tanh(\eta_-/2k_\mathrm{B}T)\sin(\eta_-\tau/\hbar)\Theta(\tau) \quad \text{for } \Delta_0 \ll J. \tag{4.64}$$

Since we have discarded an asymmetry when deriving (4.64), we have to drop Δ in (3.46). When we note the correspondence $\Delta_0 = \Delta_\mathrm{K}$ and hence $\eta = \eta_-$, we find that both expressions turn out to be identical. After inserting the projection of the dipole moment on the direction of the external field, $|p_x| = qd/2$, in (3.46), we even recover the factor 8 in (4.64). This confirms the validity of the two-state pair model for the strong-coupling case.

Low-frequency dielectric experiments are described by the Fourier transform of the response function (4.62) evaluated at zero frequency,

$$\alpha(\omega=0) = (4/\hbar)(p^2/\epsilon_0)(1+w_4)\sum_{\pm} R_{\pm}\frac{1}{\eta_{\pm}}\tanh(\beta\eta_{\pm}/2). \tag{4.65}$$

The rotary echo experiments described in the previous chapter single out impurity pairs with a well-defined coupling energy J and with a distance

4. A Pair of Two-Level Systems

vector parallel to the external field. In linear-response spectroscopy, however, one takes the sum of contributions from all tunneling systems present in the sample. Thus we have to perform the configurational average,

$$\chi(t) = (1/3V) \sum_{\text{pairs}} \alpha(t), \tag{4.66}$$

where the single-pair response function $\alpha(t)$ has been defined in (4.61) and where the sum runs over $(N/2)$ pairs, resulting in an average with respect to the density of states (4.37). The factor $\frac{1}{3}$ arises from the orientational average with respect to the external field.

When we take the Fourier transform and use (4.39), we find the dynamic susceptibility

$$\chi(\omega) = \frac{2}{3} \frac{np^2}{\epsilon_0 \Delta_0} \int dE D(E) \frac{E}{E^2 - \hbar^2 \omega^2} (1 + w_4) \tanh(\beta E/2). \tag{4.67}$$

Since most experiments are done at low frequency, we consider the limit $\omega \to 0$. When we proceed as for the specific heat, we obtain, with $x \equiv \eta_\pm/\Delta_0$ and the bounds (4.44), the zero-frequency susceptibility

$$\chi(0) = \frac{2}{3} \frac{np^2}{\epsilon_0 \Delta_0} \mu \left[\int_{x_0}^{x_-} dx + \int_{x_+}^{1/x_0} dx \right] \frac{1 + w_4}{x(1-x)^2} \tanh(x\beta\Delta_0/2). \tag{4.68}$$

According to the above discussion, the factor w_4 may take values from -1 to 1. It is proportional to the sign of J and vanishes accordingly when we assume that both signs occur with equal probability. Yet considering the results of

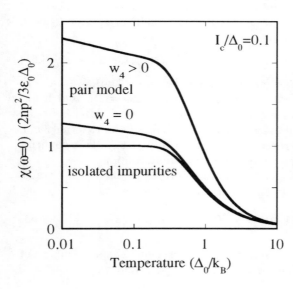

Fig. 4.5. Static susceptibility as a function of temperature for the pair model with $I_c/\Delta_0 = 0.1$. The curve labeled $w_4 > 0$ includes the interference term; it is given by (4.68). For the curve labeled $w_4 = 0$ the factor $(1 + w_4)$ has been replaced by unity. For comparison we plot the result for non-interacting impurities, (4.11)

the previous section, one is rather led to put $J > 0$ in general, corresponding to a positive w_4.

The effect of the interference term proportional to w_4 is displayed in Fig. 4.5. We plot the zero-frequency susceptibility, (4.68), both with w_4 as given in (4.31) and with $w_4 = 0$. At low temperatures the interference effect enhances the susceptibility by about 80 %, whereas it vanishes for $T \gg \Delta_0/k_\mathrm{B}$.

But even when we drop the factor $(1 + w_4)$, the susceptibility obtained from the pair model differs from that of non-interacting impurities. The low-energy excitations result in a steady increase when the temperature is lowered; the susceptibility is constant for temperatures well below the lower cut-off Δ_0^2/J_2, which corresponds to less than 10 mK. The zero-temperature value of χ is significantly larger than χ_iso.

The low-frequency susceptibility (4.68) cannot be evaluated analytically. For later use we consider the limit of zero temperature, where we have $\tanh(x\beta\Delta_0/2) = 1$ for all systems. Furthermore we drop the factor $(1+w_4)$. [It is easily restored in the integrals by inserting the relation $J = |1-x^2|x^{-1}\Delta_0$ in (4.31) and (4.33).] The remaining integrals are elementary; when we evaluate

$$\int \mathrm{d}x \frac{1}{x(1-x)^2} = \frac{1}{2}\left[\frac{1}{1-x^2} + \log\frac{x^2}{1-x^2}\right] \quad (4.69)$$

and use $x_- x_+ = 1$, we obtain, after some algebra,

$$\chi(\omega = 0) = \frac{2}{3}\frac{np^2}{\epsilon_0 \Delta_0}\mu\left[\frac{x_-^2}{1-x_-^2} + \frac{x_0^2}{1-x_0^2} + \log(x_-/x_0)\right] \text{ for } T = 0. \quad (4.70)$$

[Equation (4.70) is plotted in Fig. 5.15 on p. 111 as a function of $\mu = I_c/\Delta_0$.]

In order to form a link with the results for isolated defects, we consider the limit $\mu \to 0$; with $x_- \approx 1 - \frac{1}{2}\mu$ and $x_0 = J_2/\Delta_0$, we get

$$\chi(\omega = 0) = \frac{2}{3}\frac{np^2}{\epsilon_0 \Delta_0}\left[1 + \mu\left(\log(J_2/\Delta_0) - \frac{1}{2}\right)\right] \quad \text{for } \mu \ll 1 \quad (4.71)$$

and $T = 0$. Thus at very small concentrations, the interaction yields a correction linear in μ. For real systems the ratio J_2/Δ_0 takes values between 100 and 1000; hence the prefactor $[\log(J_2/\Delta_0) - \frac{1}{2}]$ is positive and larger than unity.

In order to render evident the limited range of validity of the pair model, we consider the case $\mu \gg 1$; from (4.44) we then find $x_- = 1/\mu \ll 1$ and after insertion in (4.70) as the leading term

$$\chi(\omega = 0) = \frac{2}{3}\frac{np^2}{\epsilon_0 \Delta_0}\mu\log(1/c) \text{ for } \mu \gg 1. \quad (4.72)$$

Note the enhancement by roughly a factor μ as compared to the susceptibility of non-interacting defects; the resulting dependence of the susceptibility on density, $\chi(\omega = 0) \propto n^2$, is obviously an unphysical result for large density or $\mu \gg 1$. (If we took the interference term into account, we would obtain an even larger value.)

5. Cross-Over to Relaxation: Mori Theory

When we augment the impurity concentration beyond a few hundred ppm, static and dynamic experiments give evidence for a significant change in the physical properties. In Fig. 5.1 we show the dielectric susceptibility as a function of temperature measured at a frequency of 10 kHz for various concentrations of ^{6}Li impurities in KCl. As already discussed in Fig. 2.4, the data for the weakly doped sample with 6 ppm ^{6}Li obey the expression for non-interacting impurities, (2.27) or (4.10). At higher concentrations, two deviations from this law become apparent.

First, the value of the susceptibility at zero temperature is longer proportional to the impurity density n, contrary to what is expected from (4.10). When we compare the data for 70 and 210 ppm, we find that a change in density by a factor of three enhances the susceptibility by merely about 50 %; a further increase to 1100 ppm results in a *decreasing* susceptibility.

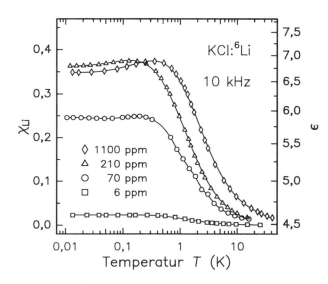

Fig. 5.1. Dielectric susceptibility of ^{6}Li as a function of temperature for various concentrations. The solid lines are guides for the eye. By courtesy of Weis and Enss [63]

Second, in addition to the so-called resonant contribution proportional to $\tanh(\Delta_0/k_B T)$, there is a relaxation peak at a temperature of about 200 mK.

Clearly the most troubling feature consists in the reduction of the susceptibility for 1100 ppm by a factor of 10 as compared with (4.10). The situation is worsened further when we note that in (4.71) the pair model predicts an enhancement with respect to the result for isolated defects. For 1100 ppm ^{6}Li, we have $\mu \approx 1$, and the correction factor in (4.71) takes a value of about 6. Yet instead of the corresponding increase by a factor 6, the observed data rather show a reduction by a factor of 10, as compared with non-interacting defects. Thus Fig. 5.1 and (4.71) prove the failure of the pair model for $c > 200$ ppm ^{6}Li. (Very similar data for ^{7}Li have been reported in [62,63].)

The pair model has been derived under the assumption that a given impurity interacts significantly at most with its nearest neighbor or, equivalently, that the average interaction energy $cJ_2 = I_c = (2np^2/3\epsilon_0\epsilon)$ is much smaller than the tunnel energy Δ_0. The upper cut-off for the dipolar energy, J_2, exceeds the two-level splitting Δ_0 by several orders of magnitude. Accordingly the condition $cJ_2 \ll \Delta_0$ is satisfied only at low concentration, c, of at the most a few hundred ppm. As $(np^2/4\pi\epsilon_0\epsilon)$ approaches Δ_0, this condition is no longer satisfied, thus requiring us to take into account the coupling to all surrounding impurities. (In a more qualitative fashion, this conclusion has already been drawn from Fig. 1.2.) In this case, the virial expansion is not a good starting point, since clusters consisting of three and more defects are not negligible and prohibit truncation at any finite order. The state of N impurities no longer factorizes, but one rather faces a N-particle problem.

Hence we need to take into account the coupling of a given impurity to all surrounding neighbors. Accordingly, we generalize the two-level pair model of the previous chapter. With the usual mapping on Pauli matrices (4.2), N defects on a lattice with N_0 sites are described by the Hamiltonian

$$H = \frac{1}{2}\sum_i \Delta_0 \sigma_x^i + \frac{1}{2}\sum_{i<j} J_{ij} \sigma_z^i \sigma_z^j, \tag{5.1}$$

where $\sigma_z^i = \pm 1$ denotes the defect at site i, dwelling in the left or right off-center position, and the labels i, j run from 1 to N. The defect concentration is given by $c \equiv N/N_0$, whereas $n = N/V$ denotes the number density of N impurities in a volume V, and $n_0 = N_0/V$ the corresponding site density; obviously one has $c = n/n_0$. (The change of sign with respect to (4.17) is of no consequence.)

The dipolar interaction energy of two impurities depends both on their distance and on the relative orientation. Simplifying the angular part by ± 1 we take

$$\frac{1}{2}J_{ij} = \pm\frac{1}{4\pi\epsilon_0\epsilon}\frac{p^2}{R_{ij}^3}, \tag{5.2}$$

which differs by an insignificant factor as compared with the corresponding expression of the pair model, (4.13). Yet there is one important difference with respect to the sign of J_{ij}. From the thorough treatment of the ground state multiplet in Chap. 3, we inferred that the dipole moments of two impurities arrange in such a way that the quantity J is positive. This is no longer justified for many dipoles on random sites, where the interaction (4.13) will take both signs with equal probability, resulting in

$$\sum_{i,j} J_{ij} = 0. \tag{5.3}$$

(This simply reflects the fact that the angular average of a dipolar energy vanishes.) Throughout this chapter we will constantly refer to the average interaction energy

$$cJ_2 = I_c = (2np^2/3\epsilon\epsilon_0), \tag{5.4}$$

which is closely related to the moments J_{ij}^n (see Appendix C).

It becomes clear that the change in the defect motion is best described in terms of the dimensionless coupling parameter

$$\mu = \frac{2}{3} \frac{np^2}{\epsilon\epsilon_0 \Delta_0}, \tag{5.5}$$

which measures the average interaction energy in units of the tunnel amplitude Δ_0. (We have already used this ratio when discussing the density of states of the pair model on p. 58.)

5.1 Continued Fraction Expansion

Equation of Motion and Correlation Function

Time evolution of any observable A is written as a linear operation on the space of quantum mechanical operators,

$$A(t) = e^{i\mathcal{L}t} A, \tag{5.6}$$

with the action of the Liouville operator given by the commutator $\hbar\mathcal{L}A \equiv [H, A]$. For later convenience we note the equations of motion

$$\hbar\dot{\sigma}_x^i = -\sigma_y^i \sum_j J_{ij}\sigma_z^j, \tag{5.7}$$

$$\hbar\dot{\sigma}_y^i = -\Delta_0 \sigma_z^i + \sigma_x^i \sum_j J_{ij}\sigma_z^j, \tag{5.8}$$

$$\hbar\dot{\sigma}_z^i = \Delta_0 \sigma_y^i. \tag{5.9}$$

(We assume $J_{ii} = 0$.) Owing to the invariance under the canonical transformation $\sigma_x^i \to \sigma_x^i$, $\sigma_y^i \to -\sigma_y^i$, $\sigma_z^i \to -\sigma_z^i$, thermal averages involving an odd number of σ_z^i and σ_y^i operators vanish, e.g., $\langle\sigma_y^i\rangle = 0 = \langle\sigma_z^i\rangle$ and

$\langle \sigma_y^i(t)\sigma_x^j(t')\rangle = 0$. (Expectation values with respect to the equilibrium state are indicated by angular brackets, $\langle ...\rangle = \text{tr}(e^{-\beta H}...)/\text{tr}(e^{-\beta H})$.)

Dynamic properties are described through symmetrized two-time correlation functions

$$(A(t)|A(t')) \equiv \tfrac{1}{2}\langle A(t)A(t') + A(t')A(t)\rangle; \tag{5.10}$$

of particular interest are correlations of the reduced coordinate σ_z^i,

$$G_i(t-t') \equiv (\sigma_z^i(t)|\sigma_z^i(t')). \tag{5.11}$$

Equation (5.10) defines an inner product in the space of dynamic observables; thus the Laplace transforms of the two-time correlation functions (5.11) can be written as matrix elements of the resolvent of the Liouville operator,

$$G_i(z) = (\sigma_z^i|[\mathcal{L}-z]^{-1}|\sigma_z^i). \tag{5.12}$$

Mori's Reduction Method

Expression (5.12) may be transformed into a continued fraction by applying the resolvent identity

$$\mathcal{P}\frac{1}{z-\mathcal{L}}\mathcal{P} = \frac{\mathcal{P}}{z-\mathcal{P}\mathcal{L}\mathcal{P} + \mathcal{P}\mathcal{L}\mathcal{Q}\dfrac{1}{z-\mathcal{Q}\mathcal{L}}\mathcal{Q}\mathcal{L}\mathcal{P}} \tag{5.13}$$

repeatedly [119]; in each step the dynamics is projected by means of

$$\mathcal{Q}_i \equiv 1 - \sum_{j=1}^{i} |A_j)(A_j| \tag{5.14}$$

and $\mathcal{P}_i \equiv 1 - |A_i)(A_i|$. Thus a set of linearly independent and normalized operators is constructed by

$$A_{i+1} = \mathcal{Q}_i \dot{A}_i (\mathcal{Q}_i \dot{A}_i | \mathcal{Q}_i \dot{A}_i)^{-1/2}, \tag{5.15}$$

with $\dot{A} = i\mathcal{L}A$. When we start with σ_z^i, the equations of motion (5.7–5.9) lead us to

$$A_1 = \sigma_z^i, \qquad A_2 = \sigma_y^i, \qquad A_3 = I_i^{-1}\sum_j J_{ij}\sigma_x^i \sigma_z^j, \tag{5.16}$$

where we have defined the constant

$$I_i^2 \equiv \sum_{j,k} J_{ij}J_{ik}\langle \sigma_z^j \sigma_z^k\rangle. \tag{5.17}$$

The first term of $\dot{\sigma}_y^i$ is proportional to σ_z^i and thus has been projected out according to (5.14).

(If not otherwise stated, we take $\hbar = 1$ in the remainder of this chapter.) We thus obtain the representation

$$G_i(z) = \cfrac{-1}{z - \cfrac{\Delta_0^2}{z - \cfrac{I_i^2}{z + N_i(z)}}}, \qquad (5.18)$$

where the memory function is given as a resolvent matrix element

$$N_i(z) = (Q_3 \dot{A}_3 | [Q_3 \mathcal{L} - z]^{-1} | Q_3 \dot{A}_3) \qquad (5.19)$$

of the force

$$\begin{aligned}Q_3 \dot{A}_3 &= \Delta_0 I_i^{-1} \sum_j J_{ij} \sigma_x^i \sigma_y^j \\ &+ I_i^{-1} \sum_{j,k \neq j} J_{ij} J_{ik} \left(\sigma_y^i (\sigma_z^j \sigma_z^k - \langle \sigma_z^j \sigma_z^k \rangle) - 2 \langle \sigma_y^j \sigma_z^k \rangle A_3 \right).\end{aligned} \qquad (5.20)$$

Since the static quantity I_i^2 can be evaluated in reasonable approximation, the difficulty of the many-body problem (5.1) has been shifted to the calculation of the memory function $N_i(z)$. We are going to evaluate $N_i(z)$ in different approximations.

In a first approach, we will truncate the continued-fraction expansion after three steps; accordingly, we rewrite (5.18) as

$$G_i(z) = -\frac{z(z + N_i(z)) - I_i^2}{(z + N_i(z))(z^2 - \Delta_0^2) - zI_i^2}. \qquad (5.21)$$

Three-Pole Approximation

Equations (5.18–5.21) are still exact; as an essential approximation, we assume $N_i(z)$ to be *(a)* a smooth function of frequency and *(b)* small as compared with the energies Δ_0 and I_i. Condition *(a)* permits us to evaluate the memory function at the poles of $G_i(z)$, and with condition *(b)* we may use the bare poles and residues. Defining the resonance

$$E_i = \sqrt{\Delta_0^2 + I_i^2}, \qquad (5.22)$$

the relaxation amplitude

$$R_i = I_i^2 / E_i^2, \qquad (5.23)$$

and the damping rates

$$\gamma_i = \frac{\Delta_0^2}{E_i^2} N_i''(0) \quad \text{and} \quad \Gamma_i = \frac{1}{2} \frac{I_i^2}{E_i^2} N_i''(E_i), \qquad (5.24)$$

we find the three-pole approximation to (5.21)

$$G_i(z) = -R_i \frac{1}{z + i\gamma} - (1 - R_i) \frac{z + i\Gamma}{(z + i\Gamma)^2 - E_i^2}. \qquad (5.25)$$

The term arising from the dipolar interaction, I_i, affects the motional spectrum in several ways. On the one hand, the resonance energy is shifted from its bare value Δ_0 to E_i and acquires an imaginary part $i\Gamma_i$, whereas, on the other hand, the corresponding residue decreases; the missing spectral weight appears at zero frequency as a quasi-elastic feature with width γ_i. For vanishing interaction, $I_i \to 0$, we find a pair of undamped poles at $\pm\Delta_0$; in the opposite limit $\Delta_0 \ll I_i$, $G_i(z)$ reduces to a Debye relaxator with rate γ_i. Evaluation of the damping rates Γ_i and γ_i is not a simple matter and we postpone it for the moment.

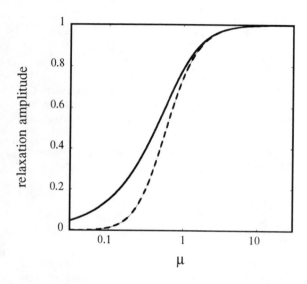

Fig. 5.2. Averaged relaxation amplitude as a function of the dimensionless coupling parameter μ. The full line gives the result of the three-pole approximation (5.27), the dashed line that of five-pole approximation (5.113)

Static properties are entirely determined by the resonance (5.22), together with the distribution function for I_i, which is derived in Appendix C. First we consider the average relaxation amplitude

$$\overline{R} = (1/N)\sum_i R_i = \int_0^\infty dI Q(I) R(I). \tag{5.26}$$

When we neglect in (5.17) correlations of spins on different sites, $\langle \sigma_z^j \sigma_z^k \rangle \to \delta_{jk}$ (for a discussion see p. 77), and use the approximate expression (C26), the integral is easily performed,

$$\overline{R} = \mu \left[\frac{\pi}{2} - \arctan(\mu) \right]. \tag{5.27}$$

Thus the relaxation amplitude vanishes in the dilute limit $\mu \to 0$, whereas it strongly increases at $\mu \approx 1$ and tends towards unity for $\mu \gg 1$ (see Fig. 5.2).

Now we turn to the dynamic susceptibility. The first term of (5.25) contributes a relaxational part

$$\chi_{\text{rel}}(\omega) = \frac{2}{3}\frac{p^2}{\epsilon_0}\frac{1}{2k_BT}\frac{1}{V}\sum_i R_i \frac{i\gamma_i}{\omega + i\gamma_i}, \tag{5.28}$$

whereas the second one gives rise to the so-called resonant susceptibility. Here we consider only the real part of the latter. Applying the Kramers-Kronig relation on the spectrum of (5.25), we find

$$\chi'_{\text{res}}(\omega) = \frac{2}{3}\frac{p^2}{\epsilon_0}\frac{1}{V}\sum_{i,\pm}(1-R_i)\frac{E_i/\hbar \pm \omega}{(E_i/\hbar \pm \omega)^2 + \Gamma_i^2}\tanh(\beta\hbar\omega/2). \tag{5.29}$$

(We reinsert $\hbar$.) The low-frequency limit $\hbar\omega \ll \Delta_0$ is of particular interest; assuming $\hbar\Gamma_i \ll E_i$, we may evaluate the temperature factor at the poles $\pm E_i$; when we insert (5.23), we thus find

$$\chi'_{\text{res}}(\omega \to 0) = \frac{2}{3V}\frac{p^2}{\epsilon_0}\sum_i \frac{\Delta_0^2}{E_i^2}\frac{1}{E_i}\tanh(\beta E_i/2). \tag{5.30}$$

The case of zero temperature turns out to be of particular interest. After inserting the distribution (C26) and $n = N/V$, we obtain

$$\chi'_{\text{res}}(0) = \frac{2}{3}\frac{np^2}{\epsilon_0}\int_{cJ_2}^{J_2} dI \frac{cJ_2}{I^2}\frac{\Delta_0^2}{(\Delta_0^2 + I^2)^{3/2}} \quad \text{at } T = 0; \tag{5.31}$$

straightforward integration yields

$$\chi'_{\text{res}}(0) = \frac{2}{3}\frac{np^2}{\epsilon_0\Delta_0}\frac{\left(\sqrt{1+\mu^2}-\mu\right)^2}{\sqrt{1+\mu^2}} \quad \text{at } T = 0. \tag{5.32}$$

In the limit of low concentrations, or $\mu \ll 1$, the resonant susceptibility is equal to that of isolated defects (4.11), whereas in the opposite limit it decreases according to

$$\chi'_{\text{res}}(0) = \begin{cases} \chi_{\text{iso}}(0) & \text{for } \mu \ll 1 \\ \chi_{\text{iso}}(0)\mu^{-3} & \text{for } \mu \gg 1 \end{cases} \quad \text{at } T = 0. \tag{5.33}$$

In physical terms, (5.27) and (5.32) both express the fact that in the dilute case $\mu \ll 1$ the motional spectrum mainly involves frequencies close to $\Delta_0/\hbar$, whereas for $\mu > 1$, the tunneling motion is governed by relaxation. Accordingly, the relaxation amplitude vanishes in the former case and the resonant susceptibility in the latter.

(We remark that the zero-temperature susceptibility of non-interacting defects (4.11), $\chi_{\text{iso}} = (2/3)(np^2/\epsilon_0\Delta_0)$, and the dimensionless parameter (5.5),

$$\mu = \frac{2}{3}\frac{np^2}{\epsilon\epsilon_0\Delta_0}, \tag{5.34}$$

differ merely by a factor ϵ; they fulfil $\mu = \chi_{\text{iso}}/\epsilon$. According to our definition, the dielectric susceptibility $\chi(\omega)$ is a dimensionless quantity.)

74   5. Cross-Over to Relaxation: Mori Theory

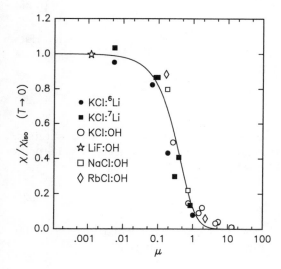

Fig. 5.3. Real part of the dielectric susceptibility for various defect systems for $T = 0$ and $\omega \to 0$ as a function of the dimensionless coupling parameter μ. Different values of μ correspond to different concentrations, according to (5.34). Thus each data point represents one sample. The solid line is given by the parameter-free curve (5.35). By courtesy of Weis and Enss [62]

At zero frequency and zero temperature, both the relaxational part and the imaginary part of χ_{res} vanish; hence we have $\chi = \chi_{\text{res}}(0)$ and with (5.32)

$$\frac{\chi}{\chi_{\text{iso}}} = \frac{\left(\sqrt{1+\mu^2} - \mu\right)^2}{\sqrt{1+\mu^2}} \quad \text{at} \quad T = 0. \tag{5.35}$$

In Fig. 5.3 this expression is plotted as a function of μ and compared with data for various defect systems. Each data point corresponds to one particular sample, the parameter μ being connected to the impurity density through (5.34).

The susceptibility (5.32) depends only on the tunnel energy Δ_0, the dipole moment p, the dielectric constant ϵ and the impurity density n. Since these quantities are known, there is no free parameter left. The only uncertainty arises from the mapping of the dipolar interaction of the actual eight-state defects, as given in (3.5), on that of the effective two-state systems, (5.2). Another choice for the relative angles would yield an additional numerical factor of the order of unity in (5.2) and in the definition of μ, resulting in a slight shift of the solid curve in Fig. 5.3. The most important experimental error involves the defect concentration; the numbers given are supposed to be correct within about 10 %. Table 5.1 gathers the parameters for the samples appearing in Fig. 5.3.

Here a word concerning our definition of the dynamic susceptibility is in order. The response function as given by (A16) is calculated with respect to the local field; accordingly, it involves the effective dipole moment $p = \epsilon_{\text{KCl}}^{-1/2} p_0$, where ϵ_{KCl} is the dielectric constant of the host crystal and p_0

Table 5.1. Parameters for the data points in Fig. 5.3. Defect density n and concentration c are related through $n = cn_0$, where, in units of 10^{22}cm^{-3}, the site density n_0 takes the values 1.61 (KCl), 6.12 (LiF), 2.23 (NaCl), and 2.81 (RbCl) [66]. After [41]

	c (ppm)		ϵ		p / D		Δ_0 / K	
KCl:^{6}Li	6, 70, 210, 1100	[62]	4.49	[67]	2.63	[69]	1.65	[69]
KCl:^{7}Li	4, 60, 215, 600	[62]	4.49	[67]	2.63	[69]	1.1	[69]
	68, 280	[37]						
KCl:OH	250, 500, 1500	[72]	4.49	[67]	1.85	[73]	0.25	[74]
	90	[37]						
	620, 1800, 4350	[37]						
	1780, 4500	[75]						
LiF:OH	300	[72]	8.50	[76]	0.089	[72]	0.87	[72]
NaCl:OH	400, 1600	[72]	5.49	[67]	1.72	[78]	1.6	[77]
RbCl:OH	68, 927	[37]	4.6^a	[79]	2.0	[73]	0.6^b	

^{a} Derived by Weis [41] from the value at room temperature $\epsilon = 4.9$ [79]
^{b} Determined from dielectric data on RbCl with 68 ppm OH [41]

the bare dipole moment of the impurity. A more common definition reads $\chi = N\alpha$, with the number of impurities N and the atomic polarizability α. This difference is most significant when considering e.g. a ferro-electric phase transition. In the present case, however, it merely results in a numerical factor for the local field correction. (See [62].)

Five-Pole Approximation

Now we go beyond the above three-pole approximation by applying the resolvent identity once more. It proves to be appropriate not to project on $Q_3\dot{A}_3$ as given in (5.20) but only on its first term, which consists of two spin operators,

$$A_4 = \tilde{I}_i^{-1} \sum_j J_{ij} \sigma_x^i \sigma_y^j, \tag{5.36}$$

with

$$\tilde{I}_i^2 = \sum_{j,k} J_{ij} J_{ik} \langle \sigma_y^j \sigma_y^k \rangle. \tag{5.37}$$

Accordingly, we split the memory function $N_i(z)$ into two parts,

$$N_i(z) = \hat{G}_i(z) + \hat{N}_i(z), \tag{5.38}$$

the first of which contains the two-mode contributions

$$\hat{G}_i(z) = \Delta_0^2 \tilde{I}_i^2 I_i^{-2} (A_4 | [\mathcal{Q}_3 \mathcal{L} - z]^{-1} | A_4), \tag{5.39}$$

and the second one the remainder. (The prefactor of $\hat{G}_i$ arises from the normalization condition $(A_4|A_4) = 1$.) When we finally project the dynamics of $\hat{G}(z)$ on A_4, we obtain with

$$M_i(z) = (\mathcal{Q}_4 \dot{A}_4 | [\mathcal{Q}_4 \mathcal{L} - z]^{-1} | \mathcal{Q}_4 \dot{A}_4), \tag{5.40}$$

the expression

$$\hat{G}_i(z) = -\frac{\Delta_0^2 \tilde{I}_i^2}{I_i^2} \frac{1}{z + M_i(z)}. \tag{5.41}$$

Both memory functions $\hat{N}_i$ and M_i involve composite operators made up of at least three spins; the two-mode contributions are contained in the static quantities I_i^2 and $\tilde{I}_i^2$ appearing in (5.18, 5.41).

Two-Mode Dynamics

It is instructive to consider the special case where the defect at the site i considered has a single neighbor at site i_0. As a consequence, time evolution of the operator σ_z^i occurs entirely in the relevant space spanned by the basis elements A_1, A_2, A_3, A_4. The memory functions M_i and $\hat{N}_i$ vanish, and the transverse correlation function takes a particularly simple form; the static correlations (5.17) and $\tilde{I}_i^2$ are equal to $I_i^2 = \tilde{I}_i^2 = J_{ii_0}^2$. From (5.18) and (5.41) we obtain in a straightforward fashion $\hat{G}_i(z) = -\Delta_0^2/z$ and

$$G_i^{II}(z) = -\frac{z(z^2 - \Delta_0^2 - I_i^2)}{(z^2 - \Delta_0^2)^2 - z^2 I_i^2}. \tag{5.42}$$

When we single out the pole contributions and use

$$\eta_\pm = \sqrt{\Delta_0^2 + \frac{1}{4} I_i^2} \pm \frac{1}{2} I_i, \tag{5.43}$$

we obtain

$$G_i^{II}(z) = \sum_\pm \frac{\eta_\mp}{\eta_+ + \eta_-} \frac{-z}{z^2 - \eta_\pm^2}. \tag{5.44}$$

Equation (5.44) agrees with the exact solution for the dynamics of a coupled pair of two-level systems given in the previous section; taking $I_i \to J$, we find that (5.44) is identical to (4.54). This coincidence is not surprising; the mapping $\sigma_\alpha^i \to \sigma_\alpha$ and $\sigma_\alpha^{i_0} \to \sigma_\alpha'$ reveals that the operators $A_1, ..., A_4$ defined above are identical to the first set of (4.19). Yet for a single neighbor i_0, the exact dynamics of σ_z^i is confined to that subspace.

5.2 Mode-Coupling Approximation

The exact equations (5.18) and (5.41) serve as starting points for an improved calculation of the correlation function $G_i(z)$. The essential approximations consist in a decoupling scheme for both static and dynamic correlations and a Markov approximation for the memory functions.

Decoupling of Static Correlations

In order to simplify the quite complicated expectation values appearing in the continued fraction representation of the resolvent matrix element (5.12), thermal averages of products of spin operators defined on different sites are replaced by appropriate products of expectation values,

$$\langle \sigma_\alpha^i \sigma_\beta^k \rangle \approx \langle \sigma_\alpha^i \rangle \langle \sigma_\beta^k \rangle \quad \text{for } i \neq k, \tag{5.45}$$

$$\langle \sigma_\alpha^i \sigma_\beta^j \sigma_\gamma^k \rangle \approx \langle \sigma_\alpha^i \rangle \langle \sigma_\beta^j \rangle \langle \sigma_\gamma^k \rangle \quad \text{for } i \neq j \neq k \neq i. \tag{5.46}$$

Since $\langle \sigma_z^i \rangle = 0 = \langle \sigma_y^i \rangle$, the quantities (5.17) and (5.37) then read

$$I_i^2 = \tilde{I}_i^2 = \sum_j J_{ij}^2; \tag{5.47}$$

the force operators defining the memory functions take the much simpler form

$$\mathcal{Q}_3 \dot{A}_3 - \Delta_0 A_4 = I_i^{-1} \sum_{j,k} J_{ij} J_{ik} \sigma_y^i \sigma_z^j \sigma_z^k \equiv F_3, \tag{5.48}$$

$$\mathcal{Q}_4 \dot{A}_4 = I_i^{-1} \sum_{j,k \neq i} J_{ij} J_{jk} \sigma_x^i \tilde{\sigma}_x^j \sigma_z^k + I_i^{-1} \sum_{j,k \neq j} J_{ij} J_{ik} \sigma_y^i \sigma_y^j \sigma_z^k \equiv F_4, \tag{5.49}$$

with the abbreviation $\tilde{\sigma}_x^j \equiv \sigma_x^j - I_i^{-1} J_{ik} \langle \sigma_x^j \rangle A_4$; the latter correction, however, is small and may be neglected. (In the sequel, contributions fulfilling $I_i \approx |J_{ij}| \gg |J_{ik}|$ prove to be most relevant; accordingly, one has $\tilde{\sigma}_x^j \approx \sigma_x^j$.)

Decoupling of the Memory Functions

The memory functions $M_i(t)$ and $\hat{N}_i(t)$ are determined by the operators F_4 and F_3, as defined in (5.48, 5.49); evaluation of the time-dependent correlations of these composite operators requires further approximations. The mode-coupling scheme involves several steps.

First, we neglect time-dependent correlations $\langle A(t) B \rangle$, which vanish in the static limit $t = 0$; this concerns mainly two-time correlations of $\mathcal{Q}_3 \dot{A}_3$ and A_4. Second, in the time evolution the projected Liouville operator is replaced by the full one, $\mathcal{L}$,

$$M_i(t) \approx (F_4 | e^{-i\mathcal{L}t} F_4), \tag{5.50}$$

5. Cross-Over to Relaxation: Mori Theory

$$\hat{N}_i(t) \approx (F_3|e^{-i\mathcal{L}t}F_3). \tag{5.51}$$

Third, correlations of the composite operators F_α are approximated by appropriate products of simpler correlation functions.

With $\dot{\sigma}_z^i = \Delta_0 \sigma_y^i$ it is clear that $\hat{N}_i(t)$ and the part of $M_i(t)$ arising from the second term in (5.49) involve only correlations of operators $\dot{\sigma}_z^i$ and σ_z^i. The first term in (5.49),

$$F_K = I_i^{-1} \sum_{j,k \neq i} J_{ij} J_{jk} \sigma_x^i \sigma_x^j \sigma_z^k, \tag{5.52}$$

gives rise to quite a different memory function; thus we are led to divide $M_i(t)$ into two parts,

$$M_i(t) = K_i(t) + \hat{M}_i(t), \tag{5.53}$$

the first one being defined by

$$K_i(t) = (F_K|e^{-i\mathcal{L}t}|F_K), \tag{5.54}$$

and the second one as the remainder.

Now we apply the mode-coupling approximation on the memory functions. We start with $\hat{M}_i(t)$ and $\hat{N}_i(t)$ and evaluate them according to the decoupling scheme

$$\langle \sigma_{\alpha_1}^{i_1}(t) \sigma_{\alpha_2}^{i_2}(t) \sigma_{\alpha_3}^{i_3}(t) \sigma_{\beta_1}^{j_1} \sigma_{\beta_2}^{j_2} \sigma_{\beta_3}^{j_3} \rangle \approx \sum_P \prod_{n=1}^3 \delta_{i_n \hat{j}_n} \delta_{\alpha_n \hat{\beta}_n} \langle \sigma_{\alpha_n}^{i_n}(t) \sigma_{\hat{\beta}_n}^{\hat{j}_n} \rangle, \tag{5.55}$$

where $\hat{j}_n \equiv j_{Pn}$ and $\hat{\beta}_n \equiv \beta_{Pn}$ run over all permutations P of the triples j_n and β_n. Operators with the same time argument commute, since the restriction on the summation label for $n \neq n'$ implies the relations $i_n \neq i_{n'}$ and $j_n \neq j_{n'}$.

When we insert (5.55) and use time translational invariance, $\langle \sigma_\alpha^j(t) \sigma_\alpha^j \rangle = \langle \sigma_\alpha^j \sigma_\alpha^j(-t) \rangle$, we obtain

$$\hat{M}_i(t) = \sum_{j,k \neq j} \frac{J_{ij}^2 J_{ik}^2}{2 I_i^2} \sum_{\pm} \langle \sigma_y^i(\pm t) \sigma_y^i \rangle \langle \sigma_y^j(\pm t) \sigma_y^j \rangle \langle \sigma_z^k(\pm t) \sigma_z^k \rangle, \tag{5.56}$$

$$\hat{N}_i(t) = \sum_{j,k \neq j} \frac{J_{ij}^2 J_{ik}^2}{I_i^2} \sum_{\pm} \langle \sigma_y^i(\pm t) \sigma_y^i \rangle \langle \sigma_z^j(\pm t) \sigma_z^j \rangle \langle \sigma_z^k(\pm t) \sigma_z^k \rangle. \tag{5.57}$$

The various contractions of the operators σ_z^j lead to an additional factor of two in $\hat{N}_i$.

Fourier transformation yields the spectral representation of the memory functions. Using (A10, A11) and the differentiation theorem for $\Delta_0^2 \langle \sigma_y^i(t) \sigma_y^i \rangle = -\partial_t^2 \langle \sigma_z^i(t) \sigma_z^i \rangle$, the spectra of $\langle \sigma_z^i(\pm t) \sigma_z^i \rangle$ and $\langle \sigma_y^i(\pm t) \sigma_y^i \rangle$ are expressed through $G_i''(\omega)$. With the weighted convolution integral (A13), one finds for the spectra of (5.56) and (5.57)

$$\hat{M}_i''(\omega) = \sum_{j,k \neq j} \frac{J_{ij}^2 J_{ik}^2}{I_i^2 \Delta_0^4} (\omega^2 G_i''(\omega) * (\omega^2 G_j''(\omega) * G_k''(\omega))), \tag{5.58}$$

$$\hat{N}_i''(\omega) = 2 \sum_{j,k \neq j} \frac{J_{ij}^2 J_{ik}^2}{I_i^2 \Delta_0^2} (\omega^2 G_i''(\omega) * (G_j''(\omega) * G_k''(\omega))). \tag{5.59}$$

(The reactive parts $\hat{M}_i'(\omega)$ and $\hat{N}_i'(\omega)$ prove to be negligible as compared with the dissipative functions $\hat{M}_i''(\omega)$ and $\hat{N}_i''(\omega)$, and thus are not considered further here.)

Through (5.58, 5.59) the memory functions $\hat{M}_i''(\omega)$ and $\hat{N}_i''(\omega)$ are determined by the correlation spectra $G_k''(\omega)$. Owing to the factors ω^2 in the convolution integrals, both $\hat{M}_i''(\omega)$ and $\hat{N}_i''(\omega)$ are smooth functions of frequency even at $\omega = 0$.

The Singular Contribution $K_i(z)$

Now we turn to the remaining term (5.54). Since it involves operators σ_x^i, it cannot be expressed through the spectra $G_k''(\omega)$. Contrary to $\hat{M}_i''(\omega)$ and $\hat{N}_i''(\omega)$, we expect the spectrum $K_i''(\omega)$ to display a zero-frequency pole. This is most obvious when noting that the expectation value $\langle \sigma_x^i \sigma_x^j \rangle$ is finite, and that $G_k(\omega)$ may contain a pole at $\omega = 0$. Thus $K_i(t)$ comprises a term proportional to $G_k(t)$, giving rise to a zero-frequency feature in $K_i''(\omega) \propto G_k''(\omega)$.

There is no unambiguous scheme for calculating $K_i(t)$; here we resort to a most simple approximation by replacing the correlation functions in (5.54) by a Debye relaxator with some average rate ξ,

$$K_i(t) = \sum_{j,k \neq i} \frac{J_{ij}^2 J_{jk}^2}{I_i^2} e^{-\xi t}. \tag{5.60}$$

In Appendix D, we investigate (5.54) in more detail by decoupling the time correlations in two factors

$$K_i^I(t) = \sum_{j,k \neq i} \frac{J_{ij}^2 J_{jk}^2}{2 I_i^2} \sum_{\pm} \langle \sigma_x^i(\pm t) \sigma_x^j(\pm t) \sigma_x^i \sigma_x^j \rangle \langle \sigma_z^k(\pm t) \sigma_z^k \rangle; \tag{5.61}$$

after deriving a continued fraction for the first one and expressing the second one through $G_k''(\omega)$, we get a set of self-consistent equations. Still another factorization scheme has been used in [60],

$$K_i^{II}(t) = \sum_{j,k \neq i} \frac{J_{ij}^2 J_{jk}^2}{2 I_i^2} \sum_{\pm} \langle \sigma_x^i(\pm t) \sigma_x^i \rangle \langle \sigma_x^j(\pm t) \sigma_x^j \rangle \langle \sigma_z^k(\pm t) \sigma_z^k \rangle, \tag{5.62}$$

which results in self-consistency relations similar to those discussed in the appendix.

80   5. Cross-Over to Relaxation: Mori Theory

Clearly these approximations lead to different results for $K_i''(\omega)$. It turns out, however, that only the low-frequency part of this spectrum is of interest; the corresponding spectral weight is given by the residue of the zero-frequency pole of $K_i(z)$. With

$$\kappa_i^2 = \sum_{j,k \neq j} \frac{J_{ij}^2 J_{jk}^2}{I_i^2} \tag{5.63}$$

one obviously has

$$K_i(z) = -\frac{\kappa_i^2}{z + i\xi}. \tag{5.64}$$

Since in general $K_i(t)$ will comprise oscillatory terms besides the relaxation contribution, the relaxation amplitude arising from (5.61) or (5.62) is expected to be somewhat smaller than κ_i^2. For our purpose, however, this difference is of little significance (see Appendix D).

5.3 Markov Approximation

In order to reduce the mode-coupling integrals to algebraic equations, we make an ansatz for the function $G_i(z)$. The continued fraction (5.18) yields two pairs of complex conjugate poles, and the function K_i will be shown to contribute a relaxational pole on the imaginary axis. Thus we are led to the ansatz

$$G(z) = -R\frac{1}{z + i\gamma} - \sum_{\pm} R_\pm \frac{z + i\Gamma_\pm}{(z + i\Gamma_\pm)^2 - E_\pm^2}. \tag{5.65}$$

In this section we derive relations that permit us to calculate the parameters $R_\pm$, $E_\pm$ and $\Gamma_\pm$ on every site. (Here and in the sequel, the site label i is suppressed when a mistake is unlikely.)

Amplitudes and Resonances

If we assume the damping constants to be much smaller than the energies $E_\pm$, we may in a first step discard damping effects and calculate the amplitudes R_α and the energies $E_\pm$ from (5.41) by neglecting the functions $\hat{N}(z)$ and $\hat{M}(z)$. Accordingly, the singular part $K(z)$ is replaced with

$$K^0(z) = -\frac{\kappa^2}{z}; \tag{5.66}$$

thus we find for the continued fraction (5.41)

$$G^0(z) = -\frac{(z + K^0(z))(z^2 - I^2) - z\Delta_0^2}{z(z + K^0(z))(z^2 - I^2 - \Delta_0^2) - (z^2 - \Delta_0^2)\Delta_0^2}. \tag{5.67}$$

When we insert (5.66) and singe out the pole contributions, we obtain
$$G^0(z) = -R\frac{1}{z} - \sum_\pm R_\pm \frac{z}{z^2 - E_\pm^2}, \tag{5.68}$$
where the residues are given by
$$R = \frac{\kappa^2 I^2}{\Delta_0^4 + \kappa^2(I^2 + \Delta_0^2)} \tag{5.69}$$
$$R_\pm = \frac{E_\pm^2(E_\pm^2 - \Delta_0^2 - I^2) - \kappa^2(E_\pm^2 - I^2)}{E_\pm^2(E_\pm^2 - E_\mp^2)}, \tag{5.70}$$
and the resonances by
$$E_\pm = \sqrt{\Delta_0^2 + \frac{1}{2}(I^2 + \kappa^2) \pm \sqrt{\Delta_0^2 I^2 + \frac{1}{4}(I^2 - \kappa^2)^2}}. \tag{5.71}$$

The sum rule for the correlation spectrum, $\int d\omega G''(\omega) = \pi$, reads in terms of the amplitudes $R + R_- + R_+ = 1$. One can easily check that for a single neighbor the quantity κ vanishes, and that $G^0(z)$ reduces to the expression (5.44).

Damping Rates

Now we are going to calculate the damping rates from the memory functions $\hat{N}(z)$ and $\hat{M}(z)$ in Markov approximation. The continued fraction (5.18, 5.41) is rewritten as a rational function, which has a denominator consisting of a polynomial of fifth order in z. When we anticipate $\gamma, \Gamma_\pm \ll E_\pm$, we retain only terms linear in $\hat{N}$ and $\hat{M}$; thus we use (5.66) instead of $K(z)$ and drop $\hat{N}\hat{M}$ in the denominator of $G(z)$,
$$z(z^2 - E_+^2)(z^2 - E_-^2)$$
$$+ (z^2 - \Delta_0^2)(z^2 - \kappa^2)\hat{N}(z) + z^2(z^2 - \Delta_0^2 - I^2)\hat{M}(z). \tag{5.72}$$

The first term yields the bare poles, whereas the remainder accounts for damping.

First we consider the relaxation rate γ. The bare pole is given by the factor z in the first term of (5.72). The Markov approximation amounts to setting $z = 0$ elsewhere. Then the prefactor of $\hat{M}(z)$ vanishes; dividing (5.72) by $E_-^2 E_+^2 = \Delta_0^4 + \kappa^2(I^2 + \Delta_0^2)$, we find
$$\gamma = \frac{\Delta_0^2 \kappa^2}{\Delta_0^4 + \kappa^2(I^2 + \Delta_0^2)} \hat{N}''(0). \tag{5.73}$$

As to the inelastic resonances $E_\pm$, the second term of (5.72) proves to be small compared with the third one. When we retain only the latter, we obtain
$$\Gamma_\pm = \frac{1}{2} \frac{E_\pm^2 - I^2 - \Delta_0^2}{E_\pm^2 - E_\mp^2} \hat{M}''(E_\pm). \tag{5.74}$$

82   5. Cross-Over to Relaxation: Mori Theory

Equations (5.69–5.71) and (5.73, 5.74) determine the resolvent function $G(z)$, (5.65).

Noting that the pole E_+ is quite similar to that derived in three-pole approximation, (5.22), reveals the two expressions (5.25) and (5.65) to differ mainly through the existence of a second pair of poles at small energies, $\pm E_-$.

Now we have expressed the parameters appearing in the ansatz (5.65) by Δ_0, κ, and I, and through the memory functions derived in the previous section. We emphasize the different physical meaning of the various parts of the memory functions. The regular parts $\hat{M}(z)$ and $\hat{N}(z)$ result in a finite width of the resonances of $G(z)$, whereas $K(z)$ leads to a relaxation pole at $z \approx 0$.

When we assume $\gamma, \Gamma_\pm \ll E_\pm$, Laplace back-transformation of (5.65) yields the time-dependent correlation function

$$G(t) = Re^{-\gamma t} + R_- \cos(E_- t/\hbar) e^{-\Gamma_- t} + R_+ \cos(E_+ t/\hbar) e^{-\Gamma_+ t}. \tag{5.75}$$

5.4 Mean-Field Approximation for κ^2

In the previous section, the non-linear integral equations (5.18, 5.38, 5.41), and (5.58–5.59) have been reduced to algebraic expressions determining the amplitudes (5.69, 5.70) and the energies (5.71); a key parameter is provided by the spectral weight κ_i^2. Although all parameters are known or obey a well-known distribution, we have to simplify further in order to derive explicit results.

The quantity κ_i^2 defined in (5.63) depends in a intricate manner on the configuration of the neighbouring impurities. Yet the main contribution to the sum over j stems from the nearest neighbor labeled by i_0. Thus we have $I_i^2 \approx J_{ii_0}^2$; the difference

$$I_i^2 - J_{ii_0}^2 = \sum_{j \neq i_0} J_{ij}^2 \equiv \tau_i^2 \tag{5.76}$$

is in most cases much smaller than I_i^2. Accordingly, the ratio $J_{ii_0}^2/I_i^2$ is close to unity, and we may replace the sum over j with the term $j = i_0$,

$$\kappa_i^2 = \sum_{k \neq i} J_{i_0 k}^2. \tag{5.77}$$

Now we assume the above statement to hold true for the opposite case, i. e., impurity i to be the nearest neighbor of site i_0 and $I_{i_0}^2 \approx J_{ii_0}^2 \gg \tau_{i_0}^2$. (There are very few configurations where this assumption is wrong.) Thus we have with the definition (5.76),

$$\kappa_i^2 = \tau_{i_0}^2. \tag{5.78}$$

Accordingly, κ_i^2 obeys the distribution (C38); its average value is given by

$$\overline{\kappa^2} = \log(1/c) I_c^2. \tag{5.79}$$

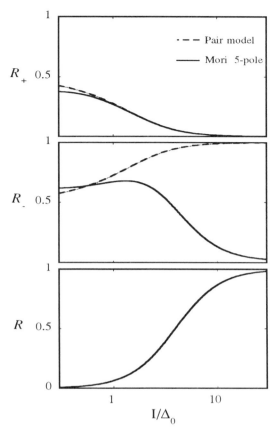

Fig. 5.4. Residues of the finite energy excitations, $R_\pm$, and of the relaxation pole, R, as a function of the dipolar interaction energy I. The value for the coupling parameter, $\mu = 1/10$, corresponds to weak coupling or low density. The variable κ^2 has been replaced by the average value, according to (5.79)

Roughly speaking, I is given by the dipolar interaction of nearest-neighbor impurities, whereas κ rather involves the coupling to the next-nearest neighbor. Because of the strong distance dependence of the dipolar interaction J_{ij}, the mean value of I is larger than that of κ. (Compare Appendix C.)

Residues

In Fig. 5.4 we compare the amplitudes obtained from the pair model and from Mori theory, depending on the dipolar interaction energy I for the particular value $\mu = \frac{1}{10}$. (We replace κ^2 with its mean value (5.79).) Note that the discrepancies for small I are of little significance, since in this range all energies are close to Δ_0, and both Mori theory and the pair model yield $R_+ + R_- = 1$.

As to the residue of the upper branch of the density of states, R_+, there is little difference between the expressions (4.32) and (5.70). Yet the amplitude of the low-energy excitations, R_-, reveals the basic modification introduced by Mori theory. Whereas the result from the pair model, (4.32), increases monotonically from $\frac{1}{2}$ to unity, Mori theory yields a strongly suppressed residue at large I. The physical origin of this reduction is displayed most clearly by the approximate formula (5.89) (which is hardly distinguishable from (5.70)). The additional factor in (5.89) is immaterial for $\eta_- \gg \kappa$ which in the case of small μ is equivalent to $I\kappa \ll \Delta_0^2$; yet for sufficiently large I the 'asymmetry energy' κ exceeds the 'tunnel energy' η and thus turns down the amplitude R_-. The missing weight in the coupled density of states reappears as relaxation amplitude $R = 1 - R_- - R_+$.

For small μ, the condition $\eta_- \approx \kappa$ corresponds to $I/\Delta_0 \approx \mu^{-1}$, in accordance with Fig. 5.4. Thus the second factor of R_- in (5.89) tends towards unity in the limit $\mu \to 0$, and we recover the results of the pair model. Yet in the opposite case $\mu > 1$, the weighted spectral density of the finite-energy resonances disappears rapidly. This is an important result, in view of the dependence of the resonant susceptibility (5.94) on the residues $R_\pm$.

5.5 Density of States

Static quantities such as the specific heat are determined by the density of states

$$\rho(E) \equiv \frac{1}{2N} \sum_i \sum_\pm \delta(E - E_\pm^{(i)}), \tag{5.80}$$

whereas the dynamic susceptibility involves the coupled density of states

$$D(E) \equiv \frac{1}{N} \sum_i \sum_\pm R_\pm^{(i)} \delta(E - E_\pm^{(i)}); \tag{5.81}$$

we note the sum rules $\int \mathrm{d}E \rho(E) = 1$ and $\int \mathrm{d}E D(E) = 1 - \overline{R}$.

The formal expressions (5.80) and (5.81) may be written in terms of the distributions for I and κ,

$$\rho(E) = \frac{1}{2} \sum_\pm \int \mathrm{d}I \int \mathrm{d}\kappa \, Q(I) Q_1(\kappa) \, \delta(E - E_\pm(I,\kappa)), \tag{5.82}$$

$$D(E) = \sum_\pm \int \mathrm{d}I \int \mathrm{d}\kappa \, Q(I) Q_1(\kappa) R_\pm(I,\kappa) \, \delta(E - E_\pm(I,\kappa)), \tag{5.83}$$

the latter being given by (C23) and (C38).

In Fig. 5.5 we plot the density of states (5.82) for several values of the coupling parameter μ. Comparison with that obtained from (4.37) for the pair model, as shown in Fig. 4.2, reveals little difference for $\mu = 0.1$. Small deviations arise mainly from the additional variable κ with distribution function (C38); the average over κ results in the exponential cut-off at very small energies and a less sharp gap about Δ_0.

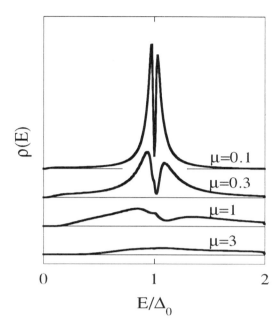

Fig. 5.5. Density of states for various values of the parameter μ as obtained from Mori theory, according to (5.82)

For $\mu \ll 1$ the density of states has equal weight below and above the value $E = \Delta_0$; the broadening of the spectrum with increasing density n, or coupling parameter μ, results in an enhancement of the low and high energy tails, as is obvious from the curves for $\mu = 0.1$ and $\mu = 0.3$. Yet for values of μ approaching and exceeding unity, we observe a transfer of spectral weight to higher energies, thus diminishing and finally suppressing low-energy excitations. All curves of Fig. 5.5 are drawn on the same scale; the plotted part of the spectra, $\int_0^{2\Delta_0} dE \rho(E)$, accounts for 97, 90, 67, and 30 % of the sum rule $\int dE \rho(E) = 1$. Thus for $\mu = 3$ about 70 % of the excitations have energies larger than $2\Delta_0$.

Now we turn to the coupled density of states, (5.81). Because of the smooth energy dependence of the residues $R_\pm$, $D(E)$ differs little from the bare density of states $\rho(E)$. In order to render obvious the novel features introduced by the present approach, as compared to the pair model, we plot in Fig. 5.6 the coupled DOS for two values of μ, corresponding to the limits of weak and strong coupling, or low and high impurity density.

For the case of low density, $\mu = 0.2$, a few minor differences of the two curves arise from the additional variable κ of Mori theory, as discussed above for $\rho(E)$. Regarding the case of higher density, $\mu = 2$, the two approaches yield basically different results. The integral of the coupled density of states obtained from Mori theory is smaller than unity, $\int dE D(E) = 1 - \overline{R}$, and

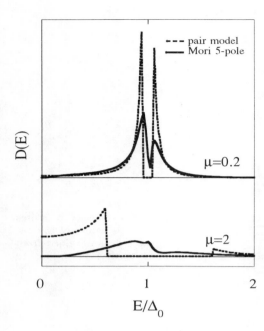

Fig. 5.6. Coupled density of states for $\mu = 0.2$ and $\mu = 2$. The full line is obtained according to (5.83); the dashed line results from the pair model, (4.39)

it vanishes at small energies. On the other hand, the pair model yields a function $D_{\rm PM}(E)$ which increases at small energy and obeys the sum rule $\int dE D_{\rm PM}(E) = 1$. (Compare the discussion on p. 65.)

Most observable quantities involve averages over the parameters I and κ. As is obvious from the intricate dependence of energy and residues, (5.70) and (5.71), the resulting integrals cannot been treated analytically. An essential simplification is achieved by using the approximate expressions

$$E_+ = \eta_+, \tag{5.84}$$

$$E_- = \sqrt{\eta_- + \kappa^2} \tag{5.85}$$

instead of (5.71) and by replacing κ^2 by its mean value $\overline{\kappa^2}$.

We shall use these approximations mainly when calculating the dielectric susceptibility. Here we briefly consider the density of states. As for the pair model it consists of two branches,

$$\rho(E) = \rho(E)_- + \rho(E)_+; \tag{5.86}$$

with (5.84) the upper one is identical to (4.37)

$$\rho_+(E) = I_c \frac{E^2 + \Delta_0^2}{(E^2 - \Delta_0^2)^2} \quad \text{for} \quad \sqrt{\Delta_0^2 + \tfrac{1}{4}I_c^2} + \tfrac{1}{2}I_c \leq E \leq J_2, \tag{5.87}$$

whereas the lower one is obtained by changing from the variable η_- to E,

$$\rho_-(E) = I_c \frac{E^2 - \kappa^2 + \Delta_0^2}{(E^2 - \kappa^2 - \Delta_0^2)^2} \frac{E}{\sqrt{E^2 - \kappa^2}}$$
$$\text{for } \Delta_0^2/J_2 \leq \sqrt{E^2 - \kappa^2} \leq \sqrt{\Delta_0^2 + \tfrac{1}{4} I_c^2} - \tfrac{1}{2} I_c. \quad (5.88)$$

In order to derive a corresponding expression for the coupled density of states $D(E)$, we have to simplify the amplitudes (5.70); using the quantities $\eta_\pm$ defined in (5.43), we approximate (5.70) as

$$R_+ = \frac{\Delta_0^2}{\Delta_0^2 + \eta_+^2}, \qquad R_- = \frac{\Delta_0^2}{\Delta_0^2 + \eta_-^2} \frac{\eta_-^2}{\eta_-^2 + \kappa^2}. \quad (5.89)$$

When we note $\eta_- \eta_+ = \Delta_0^2$ and compare with the amplitudes of the pair model (4.32), we find that the quantity κ^2 leaves R_+ unchanged and reduces R_- by a factor η_-^2/E_-^2. (Such a reduction is well-known to occur for a two-level system with asymmetry energy κ. Note the analogous prefactors in (3.46) and (6.15).)

Equations (5.86–5.88) badly account for the density of states at small energies. The sharp cut-off leads to a vanishing DOS for $E < \kappa$, whereas Fig. 5.5 reveals a much smoother reduction of low-energy states. (For $\mu = 1$ we have $\kappa \approx 2\Delta_0$; according to the numerical results shown in Fig. 5.5, the density of states is maximum at energies smaller than κ.) Yet most observable quantities F involve averages $\int dE \rho(E) F(E)$, with $F(E)$ being a smooth function of energy; thus it is of little significance whether we average properly over κ^2, or whether we insert the mean value $\overline{\kappa^2}$ everywhere.

In the remainder of this chapter we calculate the specific heat and the dynamical susceptibility and compare the results with various experimental data.

5.6 Specific Heat

The specific heat related to the finite-energy excitations $E_\pm$ reads in analogy to Sect. 4.5

$$C_V = 3k_B \frac{1}{2V} \sum_{i,\pm} \left(\frac{E_\pm^2}{2k_B T}\right)^2 \cosh(E_\pm/2k_B T)^{-2}. \quad (5.90)$$

In terms of the density of states, (5.80), this may be rewritten as

$$C_V = 3k_B \int dE \rho(E) (E^2/2k_B T)^2 \cosh(E/2k_B T)^{-2}. \quad (5.91)$$

(The factor 3 assures the proper low-density limit; see p. 52.)

The specific heat at low temperatures reflects the low-energy tail of the density of states. Thus we have to use (5.82) rather than the approximate expression (5.86). Accordingly, the specific heat cannot be evaluated analytically.

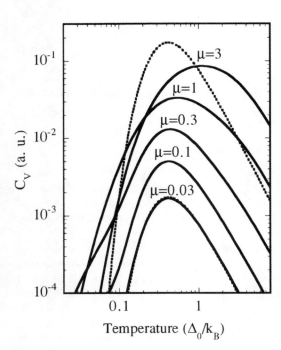

Fig. 5.7. Specific heat of interacting impurities for various values of the dimensionless coupling parameter μ. The full lines are obtained from Mori theory according to (5.91); the dashed lines give the specific heat of non-interacting impurities, (2.16), for the values $\mu = 0.03$ and $\mu = 3$

In Fig. 5.7 we plot numerical results obtained from (5.82) and (5.91) for various values of the coupling parameter μ. The specific heat displays all features of the density of states as discussed in the previous section. At very low densities, $\mu = 0.03$, (5.91) differs little from the Schottky peak of non-interacting impurities, which is displayed by the dashed line. At the opposite limit of high densities, the curves for $\mu = 3$ display a significant broadening of the specific heat, and a shift to higher temperatures.

At moderate concentration, we find for $\mu = 0.1$ an excess specific heat at low temperatures, which is very close to the result of the pair model. At somewhat higher density, $\mu \geq 0.3$, the transfer of spectral weight from small to large energies reduces the specific heat at low temperatures, as compared with the pair model. In Fig. 5.8 we plot the specific heat obtained from Mori theory for $\mu = 0.3$ with that of the pair model and with the Schottky peak of non-interacting impurities. These curves are to be compared with the data for ^{6}Li impurities shown in Fig. 4.4 on p. 60. Those for the samples with 15 ppm and 70 ppm are perfectly well fitted by the pair model. Yet the specific heat of the 140 ppm sample is smaller at low temperature than predicted by the pair model. When we note that the full line of Fig. 4.4 corresponds to the dashed one in Fig. 5.8, it is clear that Mori theory provides a better fit at low temperatures. In physical terms, this confirms the suppression of low-energy pair excitations, in accordance with the discussion in the previous section.

In Figs. 5.9 and 5.10 we show specific data for the lighter isotope ^{7}Li. Impurity concentration, impurity density, and the coupling parameter μ for each of the samples are gathered in Table 5.6. As is clear from the values for the parameter μ, these six samples cover the cross-over range from $\mu = 0.025$ up to $\mu = 1.65$. As in Fig. 5.8, the solid line gives the specific heat calculated from Mori theory, (5.91), the dashed line that of the pair model, (4.43), and the short-dashed line indicates the Schottky peak (2.16). All theoretical curves are calculated without any adjustable parameter; both the shape and the prefactor are determined by the numbers given in Table 5.6.

The data for the weakly doped samples gathered in Fig. 5.9 are similar to those obtained for ^{6}Li shown in Fig. 4.4. For the lowest concentration 17 ppm the solid line is almost identical to a single Schottky peak at a temperature of about $\Delta_0/2k_B$, in close agreement with the data. Yet the samples with 60 ppm and 203 ppm exhibit an excess specific heat at very low temperatures. The solid line for 60 ppm differs little from the result from the pair model; yet it is significantly broader than the Schottky curve. As to the sample containing 203 ppm ^{7}Li, we have plotted all three curves; the Schottky peak is too narrow, whereas the pair model overestimates the density of states at low energy. Mori theory lies between both and provides the best fit to the

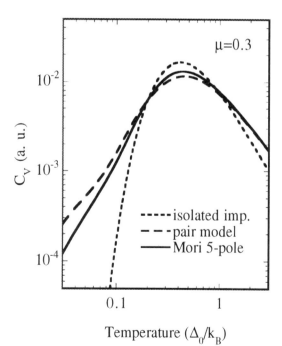

Fig. 5.8. Comparison of the specific heat as obtained from Mori theory, (5.91), from the pair model, (4.43), and for non-interacting impurities, (2.16), for the value $\mu = 0.3$

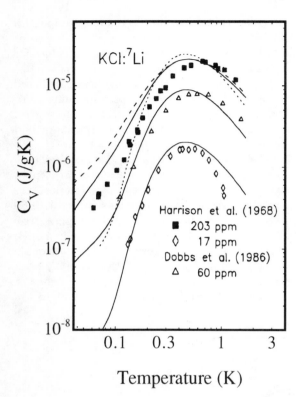

Fig. 5.9. Specific heat of ^{7}Li impurities at three different concentrations. The solid lines are calculated from Mori theory, according to (5.91). For the 203 ppm sample we have indicated the result from the pair model, (4.43), by the dashed line, and the Schottky curve, (2.16), by the short-dashed line. For the values of the coupling parameter μ see table 5.6. Data from [45,57] after Weis [41]

data. (Note the logarithmic scale; the solid line is in fact significantly closer to the data points.)

The broadening of the specific heat as a function of temperature, shown in Figs. 4.4 and 5.9, gives evidence for the existence of low-energy excitations, as discussed on p. 60. Despite the somewhat too large values at very low temperatures, the pair model would seem to account for the experimental findings.

This picture, however, changes completely when we consider the data for the samples with higher impurity density in Fig. 5.10. Although the concentration c varies by more than a factor of three from 364 ppm to 1180 ppm, the specific heat below 0.3 K turns out to be roughly the same for the three samples. In terms of the density of states this means that for concentrations well above 200 ppm the relative spectral weight at low energies decreases, and that adding impurities mainly creates states with energies at 1 K and above. Thus this feature supports the discussion of the density of states in the previous section.

The theoretical curves obtained from (5.91) account reasonably well for the data; the relative error does not exceed a factor of two. Yet a much

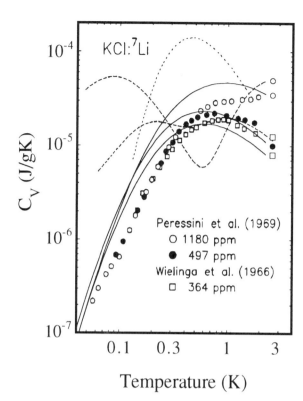

Fig. 5.10. Specific heat of ^{7}Li impurities in KCl at three different concentrations. The solid lines are given by Mori theory, (5.91). The dashed lines indicate the specific heat calculated for the pair model, (4.43), for 364 ppm and 1180 ppm, and the short-dashed line represents the Schottky peak, (2.16), for 1180 ppm. For the values of the coupling parameter μ see Table 5.6. Data from [54,58] after Weis [41]

better agreement is observed for the variation with concentration. Like the data points, the three curves are hardly distinguishable at low temperatures, $T < 200$ mK. At higher temperatures, e.g., above 1 K, they scale with the impurity density, again in accordance with the observed data.

In order to permit a thorough comparison, we plot the curves obtained from the pair model for 364 ppm and 1180 ppm. Both are definitely incompatible with the data; at low temperatures the calculated specific heat for 1180 ppm is larger by two orders of magnitudes than the measured values. This unphysical low-temperature feature has the same origin as the ever-increasing susceptibility (4.72); both arise from the low-energy peak in the density of states. (Compare Fig. 4.2 on p. 57.) For the 1180 ppm sample we show in addition the Schottky curve; at $T = 0.5$ K it is much too large, whereas it misses the wings at low and high temperatures.

In summary, Fig. 5.10 supports the discussion of the density of states of the previous section. The qualitative change at $\mu \approx 1$, displayed by Fig. 5.10, is intimately connected to the cross-over shown in Fig. 5.3 for the low-frequency susceptibility.

Table 5.2. Parameters for the specific heat data of Figs. 5.9 and 5.10. The site density is given by the density of potassium ions in a pure crystal, $n_0 = 1.61 \times 10^{22}\,\mathrm{cm}^{-3}$. Note $n = cn_0$.

Density n / 10^{17} cm^{-3}	2.8	9.7	32.6	59	80	190
Concentration c / ppm	17	60	203	364	497	1180
μ	0.024	0.084	0.28	0.51	0.70	1.65

5.7 Resonant Susceptibility

We reinsert $\hbar$. The imaginary part of the dynamic susceptibility is obtained from the correlation spectrum (5.65) through a fluctuation-dissipation theorem,

$$\chi''(\omega) = \frac{1}{\hbar}\frac{2p^2}{3\epsilon_0}\tanh(\beta\hbar\omega/2)\frac{1}{V}\sum_i G_i''(\omega), \tag{5.92}$$

and its real part by the Kramers-Kronig relation (A7). It is found to be convenient to separate the susceptibility into two parts, according to the two terms of (5.65),

$$\chi(\omega) = \chi_\mathrm{res}(\omega) + \chi_\mathrm{rel}(\omega); \tag{5.93}$$

the quasi-elastic feature at $\omega = 0$ gives rise to relaxation contribution $\chi_\mathrm{rel}(\omega)$, whereas the finite-energy excitations at $\hbar\omega = \pm E_-$ and $\hbar\omega = \pm E_+$ result in the so-called resonant part $\chi_\mathrm{res}(\omega)$.

We start with the resonant part of the susceptibility. For $\hbar\Gamma_\pm \ll E_\pm$, we may replace the frequency in the tanh function by the resonance $E_\pm$,

$$\chi_\mathrm{res}(\omega) = \frac{2p^2}{3\epsilon_0}\frac{1}{V}\sum_{i,\pm} R_\pm \frac{E_\pm}{\hbar^2(\omega + i\Gamma_\pm)^2 - E_\pm^2}\tanh(E_\pm/2k_\mathrm{B}T). \tag{5.94}$$

Dielectric Absorption at High Frequencies

The imaginary part of the resonant susceptibility is determined by the coupled density of states,

$$\chi''_\mathrm{res}(\omega) = \frac{1}{3}\frac{np^2}{\epsilon_0}\pi D(|\hbar\omega|)\tanh(\hbar\omega/2k_\mathrm{B}T), \tag{5.95}$$

multiplied by a characteristic temperature factor.

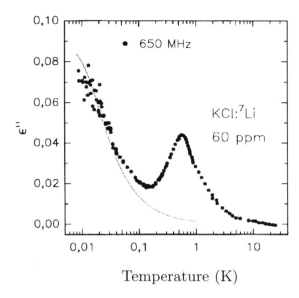

Fig. 5.11. Dielectric loss of 60 ppm ^{7}Li as a function of temperature at 650 MHz. At low temperature the resonant contribution is dominant; there is a relaxation peak at about 1 K. The solid line is given by (5.97). By courtesy of Weis and Enss [64]

According to Fig. 5.6, for $\mu \ll 1$ the coupled density of states $D(E)$ differs little from the expression of the pair model (4.39). Thus we find for low-energy pair excitations

$$D(E) = \mu \Delta_0^{-1} \quad \text{for } \Delta_0^2/J_2 \ll E \ll \Delta_0 \text{ and } \mu \ll 1. \tag{5.96}$$

When we use the definition of μ, we reach the simple result for the dielectric loss

$$\chi''_{\text{res}}(\omega) = (\pi/2)\epsilon\mu^2 \tanh(\hbar\omega/2k_\text{B}T), \tag{5.97}$$

for an external frequency satisfying $\Delta_0^2/J_2 \ll \hbar\omega \ll \Delta_0$ and $\mu \ll 1$. The prefactor in (5.97) does not depend on the resonance frequency.

Weis and Enss have measured the dielectric loss due to the resonant interaction and compared with theory [64]. Figure 5.11 shows absorption data for 60 ppm ^{7}Li impurities at $\omega/2\pi = 650$ MHz. The solid line has been obtained from a fit according to (5.97). (The applied frequency corresponds to a temperature $\hbar\omega/k_\text{B}$ of about 30 mK, thus permitting the use of the simpler expression (5.97).) There are three features worth mentioning.

(a) The temperature dependence of the data below 100 mK is a clear sign of a resonant absorption process. (An even more convincing argument is provided by experimental observation of saturation at sufficiently high field amplitude.)

(b) After Lorentz' correction for the local field [62], and with $\epsilon = 4.49$ and $\mu = 0.084$ from Table 5.6, the theoretical prefactor in (5.97) is found to be close to the value used in Fig. 5.11, thus confirming quantitatively the theoretical low-energy density of states.

(c) Comparison with similar data on 70 ppm ^{6}Li impurities reveals an isotope dependence which perfectly agrees with (5.97) and $\mu \propto \Delta_0^{-1}$ [64].

Two conclusions may be drawn from Fig. 5.11 and (5.97). First, the mere observation of the resonant absorption proves the existence of low-energy excitations in weakly doped KCl:Li. We remark that the applied frequency corresponds to an energy $E/\Delta_0 \approx 1/30$ in Fig. 5.5 on p. 85. Second, the quantitative agreement with this theory confirms the validity of our description; in particular it permits us to conclude that the assumption of a statistical distribution of the impurities is valid even for impurity pairs on nearby sites.

At higher temperatures, the dielectric loss is dominated by the relaxation contribution, which we shall consider in the next section. At low frequency, e.g., in the range of a few kHz or below, the density of states $D(E)$ vanishes, and therefore there is no resonant absorption; as a result the imaginary part of the susceptibility is determined by the relaxational contribution at all temperatures. (Compare Fig. 5.13 on p. 103.)

Reactive Part of the Susceptibility at Low Frequency

Thus at very low frequencies only the real part of the resonant susceptibility is finite, and we have for $\omega \to 0$

$$\chi_{\text{res}}(\omega = 0) = \frac{2}{3}\frac{np^2}{\epsilon_0}\frac{1}{N}\sum_{i,\pm} R_\pm E_\pm^{-1} \tanh(E_\pm/2k_\text{B}T). \tag{5.98}$$

The zero-temperature limit of this expression is of particular interest. In order to proceed further, we have to use the simplified expressions for energies and amplitudes, (5.84, 5.85) and (5.89);

We calculate separately the two terms in brackets in (5.98). The term with the plus sign is identical to that found for the pair model, (4.69); when we use (4.44),

$$x_0 = \Delta_0/J_2, \qquad x_\pm = \sqrt{1 + \frac{1}{4}\mu^2} \pm \frac{1}{2}\mu, \tag{5.99}$$

and neglect small terms of the order of x_0, we find

$$\frac{1}{N}\sum_i R_+ \frac{1}{E_+} = \frac{\mu}{2\Delta_0}\left[\frac{1}{x_+^2 - 1} + \log(1 - x_+^{-2})\right]. \tag{5.100}$$

In the dilute limit $\mu \ll 1$, this expression tends towards $(1/2\Delta_0)$. In the opposite case we have $x_+ \gg 1$ and hence $\log(1 - x_+^{-2}) \approx x_+^{-2}$; one finds that the integral vanishes as $(1/2\Delta_0\mu^3)$ for $\mu \gg 1$. We note the behavior for small and large μ

$$\frac{1}{N}\sum_i R_+ \frac{1}{E_+} = \begin{cases} 1/2\Delta_0 & \text{for } \mu \ll 1 \\ (1/2\Delta_0)\mu^{-3} & \text{for } \mu \gg 1, \end{cases} \tag{5.101}$$

and we remark that the same behavior is shown by the corresponding result in three-pole approximation, (5.32).

As a result of the additional term κ, the contribution with the minus sign proves to be quite different from that of the pair model, and rather more complicated,

$$\frac{1}{N}\sum_i R_-\frac{1}{E_-} = \frac{\mu}{\Delta_0}\int_{\Delta_0/J_2}^{x_-} \mathrm{d}x \frac{1}{x(1-x)^2}\frac{1}{(1+x^{-2}(\kappa/\Delta_0)^2)^{3/2}}. \tag{5.102}$$

We resort to a most simple approximation by replacing the last factor with an appropriate constant; we take $\kappa^2 = \log(1/c)I_c^2$ and $x^{-2} \to x_-^{-2}$, and thus have

$$\frac{1}{(1+x^{-2}(\kappa/\Delta_0)^2)^{3/2}} \to \frac{1}{(1+x_-^{-2}(\kappa/\Delta_0)^2)^{3/2}} \equiv \mathcal{F}. \tag{5.103}$$

The remaining integral is given by (4.69), and yields

$$\frac{1}{N}\sum_i \frac{R_-}{E_-} = \frac{\mu}{2\Delta_0}\left[\frac{x_-^2}{1-x_-^2} + \log\left(\frac{1-x_0^{-2}}{1-x_-^{-2}}\right)\right]\mathcal{F}. \tag{5.104}$$

This approximate result constitutes an upper bound to the integral (5.102); thus improving the integration scheme will further reduce the final value.

(By analyzing (5.102) in more detail, we have split the integral at $x' = \kappa/\Delta_0$ and replaced $\sqrt{1+x^{-2}\kappa^2/\Delta_0^2}$ with unity for $x > \kappa/\Delta_0$ and with $x^{-1}\kappa/\Delta_0$ for $x < \kappa/\Delta_0$. These integrals can be calculated exactly, providing quite cumbersome expressions; they converge, however, at the above results for small and large μ and show insignificant deviations in the intermediate range.)

For $\mu \gg 1$ we have $x_- = 1/\mu$; when we insert this in (5.104) and note $\mu = cJ_2/\Delta_0$, we obtain for the logarithm the value $2\log(1/c)$, and for the limiting cases we obtain

$$\frac{1}{N}\sum_i R_-\frac{1}{E_-} = \begin{cases} 1/2\Delta_0 & \text{for } \mu \ll 1 \\ (1/\Delta_0)\log(1/c)^{-1/2}\mu^{-5} & \text{for } \mu \gg 1 \end{cases}. \tag{5.105}$$

By inserting (5.100) and (5.104) in (5.98), taking $T = 0$, and dropping small corrections of the order x_0, we finally obtain at $\omega = 0$

$$\chi_{\text{res}} = \frac{2}{3}\frac{np^2}{\epsilon_0}\frac{\mu}{2}\left[\frac{\mathcal{F}+x_-^2}{1-x_-^2} + (1-\mathcal{F})\log(1-x_-^2) + \mathcal{F}\log(x_-^2)\right]. \tag{5.106}$$

In the dilute case $\mu \to 0$ we recover the result of the pair model. Yet at higher concentrations (μ of the order of unity or larger) we find a totally different behavior. The weak logarithmic divergence present in (4.71) is strongly suppressed by a factor μ^{-6}, which may be traced back to the last factor of (5.102). When we rewrite it as

$$\frac{\eta_-^3}{(\eta_-^2 + \kappa^2)^{3/2}}, \tag{5.107}$$

it proves obvious how that reduction occurs. Note that for large μ we have $\eta_- \leq \Delta_0^2/\mu$ and $\kappa^2 = \log(1/c)\mu^2\Delta_0^2$. When we collect these factors, we reach

96   5. Cross-Over to Relaxation: Mori Theory

the μ^{-5} law exhibited in (5.105). Clearly the evaluation of the above integral is by no means rigorous; yet this is of little significance since in any case it is much smaller than the contribution of the upper branch, (5.89).

Finally we note the low-frequency susceptibility (5.106) for the limits of low and high density,

$$\chi_{\rm res} = \chi_{\rm iso} \times \begin{cases} 1 & \text{for } \mu \ll 1 \\ \mu^{-3}\left(\tfrac{1}{2} + \mu^{-2}\log(1/c)^{-1/2}\right) & \text{for } \mu \gg 1. \end{cases} \quad (5.108)$$

In the dilute limit we recover the result for non-interacting two-level systems which increases linearly with concentration; the susceptibility exhibits a maximum at $\mu \approx 1$, i. e. $c \approx \Delta_0/J_2$, and vanishes at still higher concentrations, such as μ^{-2} or, equivalently, c^{-2}. (Note that both n and μ are proportional to the concentration c.)

Comparison with the result from three-pole approximation, (5.33), reveals an insignificant numerical difference in the limit $\mu \gg 1$. In Fig. 5.15 on p. 111 we compare (5.106) with the result derived in 3-pole approximation, (5.32). Despite of the sharper cross-over of the former, both expressions are in close agreement and confirm the consistency of the various approximations made when deriving them.

5.8 Relaxational Susceptibility

The existence of a quasi-elastic contribution to the motional spectrum (5.65) is a main result of this chapter. The resulting relaxation contribution to the dielectric susceptibility,

$$\chi_{\rm rel}(\omega) = \frac{1}{3V}\frac{p^2}{\epsilon_0}\frac{1}{k_{\rm B}T}\sum_i R_i \frac{{\rm i}\gamma_i}{\omega + {\rm i}\gamma_i}, \quad (5.109)$$

shows a complicated frequency and temperature dependence which is basically determined by the distribution for the relaxation rate γ, which in turn is an intricate function of the correlation spectra, according to (5.59) and (5.73). In this section we discuss some special features and limiting cases.

Average Relaxation Amplitude

We start with the relative weight of the relaxational part. The amplitude (5.69) depends on both I and κ. In the low-density limit we may drop $\kappa^2 \Delta_0^2$ in the denominator and find

$$R = \frac{\kappa^2}{(\Delta_0^2/I)^2 + \kappa^2} \qquad \text{for } \mu \ll 1; \quad (5.110)$$

the amplitude is most significant for strongly interacting defects whose effective tunnel energy $\eta_- = \Delta_0^2/I$ is equal to or smaller than κ. In the opposite case of high density we obtain with $\kappa \gg \Delta_0$

$$R = \frac{I^2}{\Delta_0^2 + I^2} \quad \text{for } \mu \gg 1, \tag{5.111}$$

which is identical to the result of three-pole approximation, (5.23).

Calculation of the mean value $\overline{R}$ amounts to average (5.69) over both κ^2 and I^2. As above we replace κ^2 with its average value (5.79) and thus obtain

$$\overline{R} = \int dI \tilde{Q}(I) \frac{\kappa^2 I^2}{\Delta_0^4 + \kappa^2(I^2 + \Delta_0^2)}. \tag{5.112}$$

When we insert the distribution function (C26), we can carry out a straightforward integration,

$$\overline{R} = \frac{\mu}{\sqrt{1 + \mu^{-2}\log(1/c)^{-1}}} \left[\frac{\pi}{2} - \arctan\left(\frac{\mu}{\sqrt{1 + \mu^{-2}\log(1/c)^{-1}}} \right) \right]. \tag{5.113}$$

Comparison with the result obtained from three-pole approximation, (5.27), reveals that both have the same analytical form, where in (5.113) μ is substituted with $\mu[1+\mu^{-2}\log(1/c)^{-1}]^{-1/2}$. According to Fig. 5.2 on p. 72 they show a similar cross-over from zero to unity at $\mu \approx 1$; for $\mu \gg \log(1/c)^{-1}$ both functions are identical, whereas for very small values of μ, (5.113) decreases faster to zero.

The relaxation rate (5.73) is a complicated function of the correlation spectra $G''(\omega)$. In the sequel we discuss different cases with respect to the parameter μ.

Relaxation Rate in the Dilute Case $\mu \ll 1$

Regarding the convolution integral (5.59), most neighbors fulfil $R \ll 1$ and $R_\pm \approx \frac{1}{2}$. Thus $\hat{N}''(0)$ is given by the finite-energy parts of the correlation spectra $G''_k(\omega)$; its temperature dependence is determined by the weight factors (A14) in the convolution (A13). With $\Gamma_\pm \ll E_\pm$ we obtain

$$\hat{N}''_i(0) = \frac{\pi}{\hbar} \sum_{j,k} \frac{J_{ij}^2 J_{ik}^2}{2\Delta_0^2 I_i^2} \sum_{\{\alpha,\lambda\}} E_{\alpha_i}^{(i)2} R_{\alpha_i}^{(i)} R_{\alpha_j}^{(j)} R_{\alpha_k}^{(k)}$$
$$\times \delta(E_{\alpha_i}^{(i)} + \lambda_j E_{\alpha_j}^{(j)} + \lambda_k E_{\alpha_k}^{(k)}) c_i^{-1} c_j^{-1} c_k^{-1}, \tag{5.114}$$

where the labels take the values $\alpha_j, \lambda_j = \pm$ and where we have used the shorthand notation $c_i \equiv \cosh(E_{\alpha_i}^{(i)}/2k_BT)$ for the temperature factors arising from the convolution integrals.

There is no way of evaluating (5.114) as it stands. Since we are mainly interested in its dependence on temperature and tunnel energy, we split off the factor involving the dipolar coupling,

$$J_{ik}^2 J_{ij}^2 / I_i^2 \to f_i^2. \tag{5.115}$$

The remaining sum may be rewritten as

98   5. Cross-Over to Relaxation: Mori Theory

$$\sum_{j,k} R^{(j)}_{\alpha_j} R^{(k)}_{\alpha_k} \delta(...) = \int dE_j D(E_j) \int dE_k D(E_k) \delta(...). \qquad (5.116)$$

Formally one could define f^2 in such a way that $f^2 E^{(i)2}_{\alpha_i} \Delta_0^{-2}$ times (5.116) yields again $N''_i(0)$. This is of little practical use, however. Thus we will consider f^2 as a parameter.

In the sequel we assume low temperatures $k_B T \leq \Delta_0$ and thus confine (5.114) to the low-energy excitations with $\alpha_i = -1$. When evaluating the relaxation rate, we have to distinguish several cases with respect to the ratios of E_-, Δ_0, and $k_B T$.

We start with the case of strongly coupled impurities corresponding to small energies $E_- \ll \Delta_0$. The rate consists of two contributions

$$\gamma = \gamma_P + \gamma_A, \qquad (5.117)$$

the first one arising from coupling to neighbors j,k with energies $E^{(j)}_\pm, E^{(k)}_\pm$ close to E_- and the second with energies close to the bare tunnel amplitude Δ_0. We distinguish the cases $E_- < k_B T$ and $k_B T < E_-$.

$E_- < k_B T \ll \Delta_0$. The first term γ_P stems from coupling to other impurities j,k with small energies fulfilling $E^{(i)}_- = |E^{(j)}_- \pm E^{(k)}_-|$ and $E^{(j)}_-, E^{(k)}_- \leq k_B T$. Thus the integrals in (5.116) have to be cut at $k_B T$ and yield for a smooth density of states the approximate expression

$$\hat{N}''_i(0) = \pi (f^2/\Delta_0^2) R_- E^2_- D(k_B T)^2 k_B T.$$

After insertion in (5.73) and on noting $R_- E^2_- = \eta^2_-$, this rate reads

$$\gamma_P = \frac{\pi}{\hbar} \frac{f^2 \kappa^2}{\Delta_0^4 + \kappa^2(I^2 + \Delta_0^2)} \eta^2_- D(k_B T)^2 k_B T \qquad \text{for } E_- < k_B T. \qquad (5.118)$$

The second contribution γ_A arises from neighbors j,k whose energies $E^{(j)}_\pm, E^{(k)}_\pm$ are of the order Δ_0; when we evaluate the temperature factor at Δ_0 and take $\int dE D(E) D(E - E^{(i)}_-) \equiv D_0$, we find

$$\gamma_A = \frac{\pi}{\hbar} \frac{f^2 \kappa^2}{\Delta_0^4 + \kappa^2(I^2 + \Delta_0^2)} \eta^2_- D_0 \cosh(\Delta_0/2k_B T)^{-2}. \qquad (5.119)$$

$k_B T < E_-$. Regarding systems with E_- larger than the thermal energy $k_B T$, the argument runs along the same lines as above. Yet γ_P is determined by neighbors j,k with $E^{(i)}_- = E^{(j)}_- + E^{(k)}_-$; accordingly, the integral (5.116) is cut at $E^{(i)}_-$ and yields roughly $E^{(i)}_- D(E^{(i)}_-)$, and the rate carries an appropriate temperature factor,

$$\gamma_P = \frac{\pi}{\hbar} \frac{f^2 \kappa^2}{\Delta_0^4 + \kappa^2(I^2 + \Delta_0^2)} \frac{\eta^2_- E_- D(E_-)^2}{\cosh(E_-/2k_B T)^2} \qquad \text{for } k_B T < E_-. \qquad (5.120)$$

The second term γ_A is identical to (5.119).

Equations (5.118) and (5.120) match at $k_B T = E_-$. Finally we remark that the density of states at low energies,

$$D(E) = \mu/\Delta_0 \quad \text{for } \kappa < E \ll \Delta_0, \tag{5.121}$$

is constant and much smaller than at $E \approx \Delta_0$. (Compare Fig. 5.6 on p. 86.) The rate γ_P stems from contributions with small densities of states but large temperature factors; γ_A involves a large density of states but an exponentially small temperature factor. Due to its power law behavior, γ_P will prevail at very low temperatures, whereas the activated rate γ_A is dominant as the thermal energy approaches the tunnel amplitude Δ_0.

Up to now we have considered small energy $E_- \ll \Delta_0$. Yet at low density, or $\mu \ll 1$, most impurities exhibit excitations close to the tunnel energy Δ_0. (Compare the density of states shown in Fig. 5.5 on p. 85.) For these systems the contribution γ_P, as given by (5.120), basically reduces to (5.119).

The activated rate does not depend on the energy E_-, whereas γ_P varies with both η_- and E_-. The latter rate is largest for systems fulfilling $E_- \approx k_B T$. Except for very low temperatures, the first factor in (5.118) may be replaced by $(f^2 \kappa^2 / \Delta_0^4)$, and η_- by E_-; we thus obtain the upper bound for γ_P

$$\gamma_{\text{th}} = \frac{\pi}{\hbar} \frac{f^2 \kappa^2}{\Delta_0^4} D(k_B T)^2 (k_B T)^3. \tag{5.122}$$

Equations (5.118–5.122) exhibit some features worthy of being discussed in more detail.

(a) Isotope effect. The dependence of the rates on the tunneling amplitude Δ_0 results in an unusual isotope effect; we discuss the dependence on Δ_0 of the various factors. For low density, most systems fulfil $\kappa I \ll \Delta_0^2$, resulting in a 0prefactor $f^2 \kappa^2 / \Delta_0^4$ for both γ_P and the activated rate γ_A. The second factor η_-^2 is given by $\eta_- \approx \Delta_0^2 / I$ for $\Delta_0 \ll I$ and by $\eta_- \approx \Delta_0$ for $\Delta_0 \gg I$. As to the density of states, (5.121) yields $D = I_c / \Delta_0^2$ for γ_P; for the effective value in γ_A, one has $D_0 \approx \Delta_0^{-1}$.

In the activated rate γ_A, we may replace η_-^2 by its most probable value Δ_0^2, resulting in a variation with the inverse cube of Δ_0,

$$\gamma_A = \frac{\pi}{\hbar} \frac{f^2 \kappa^2}{\Delta_0^3} \cosh(\Delta_0 / 2 k_B T)^{-2}. \tag{5.123}$$

For the remaining part γ_P, the isotope effect depends on the ratio of κ, η_-, and $k_B T$. When we take $\eta_- = \Delta_0^2 / I$ and note $D(E) = I_c / \Delta_0^2$, we find for (5.118)

$$\gamma_P = \frac{\pi}{\hbar} \frac{f^2 \kappa^2}{\Delta_0^4} \frac{I_c^2}{I^2} k_B T. \tag{5.124}$$

This behavior is valid except for very low temperatures $k_B T \ll \kappa$, where the rate γ_P rapidly vanishes. The variation of the rates (5.123) and (5.124) with the tunnel energy Δ_0 concerns the most relevant systems; considering different parameters would yield exponents ranging from –2 to –6 without changing the qualitative behavior.

This inverse isotope effect basically differs from what is usually observed for coupling to thermal lattice vibrations; there the damping rate proves to be proportional to Δ_0^2 [cf.(7.32)]. In an investigation of pairs of two-level systems with finite asymmetry energy, the phonon induced rate was found to vary with the fourth power of the tunnel energy, $\gamma_{\text{ph}} \propto \Delta_0^4$, as is expected when inserting the effective tunnel energy $\eta_- = \Delta_0^2/J$ in the well-known first-order expression for the rate [53].

By comparing the mean relaxation rate obtained for the samples with 60 ppm ^{7}Li and 70 ppm ^{6}Li, Weis and Enss found the rate for heavier isotope ^{6}Li to be significantly smaller than that for ^{7}Li. For different temperatures, they found that the ratio $^7\gamma/^6\gamma$ took values between 4 and 10 [65]. Since for phonon relaxation one expects the inverse isotope dependence, these data provide clear evidence that relaxation of lithium impurities is driven by interaction.

(b) Temperature dependence. The behavior to be observed in experiments depends very much on the applied frequency ω since the relaxation susceptibility singles out systems satisfying $\gamma \approx \omega$. At temperatures comparable to the tunneling amplitude Δ_0, the rate γ_A governs the relaxation dynamics. When decreasing the temperature well below Δ_0, (5.119) shows an Arrhenius behavior with activation energy Δ_0,

$$\gamma_A \equiv \Gamma_0 e^{-\Delta_0/k_B T} \qquad \text{for } k_B T \ll \Delta_0, \tag{5.125}$$

with a temperature-independent prefactor Γ_0.

When lowering the temperature further, the contribution γ_P will overtake the activated part γ_A. Depending on the ratio of E_- and $k_B T$, the former is either linear in temperature as in (5.118), or shows Arrhenius type behavior with activation energy E_-, according to (5.120).

Relaxational Susceptibility in the Dilute Case $\mu \ll 1$

In order to calculate the susceptibility (5.109), we would need to know the distribution function for R and γ, which is beyond the reach of our approach. Instead we replace the relaxation amplitudes R_i by the mean value and use a simplified expression for the rates. With (5.113) the average amplitude reads for small μ

$$\overline{R} = \frac{\pi}{2}\mu^2 = \frac{\pi}{2}\log(1/c)\left(\frac{2np^2}{3\epsilon\epsilon_0 \Delta_0}\right)^2 \qquad \text{for } \mu \ll 1; \tag{5.126}$$

because of its smaller tunnel energy, we expect a more pronounced relaxation peak for the heavier isotope ^{7}Li.

The difficulty arises from the distribution of the rate. We start with a qualitative discussion of the temperature dependence.

The prefactor of (5.109) varies with the inverse of the temperature, which accounts for the classical high-temperature behavior. At the opposite limit of zero temperature, the susceptibility vanishes because of $\gamma \to 0$ for $T \to 0$.

Thus both real and imaginary parts of χ_{rel} exhibit a relaxation peak which is found to occur at a temperature where the rate is roughly equal to the external frequency, $\gamma \approx \omega$; in most cases γ_A is the relevant rate, resulting in a peak temperature

$$k_B T^* = \frac{\Delta_0}{\log(\Gamma_0/\omega)}. \tag{5.127}$$

The Arrhenius law (5.125) gives rise to the weak logarithmic dependence on frequency ω and prefactor Γ_0; hence even for a wide distribution of the latter, one expects a pronounced relaxation peak that is mainly determined by the activation energy Δ_0.

For sufficiently large f^2, the condition $\gamma \approx \omega$ is only satisfied at lower temperatures, where the power law contribution γ_P exceeds the activated rate. The parameters f^2 and η_- are not the same for all impurities but rather cover a broad range. According to the linear temperature dependence of γ_P, there are systems fulfilling $\omega \approx \gamma_P$ at any temperature $T < T^*$.

Thus the relaxational susceptibility consists of two features, a peak at temperature T^* and a washed-out contribution at lower temperatures.

Regarding the frequency dependence, the imaginary part $\chi''_{\text{rel}}(\omega)$ shows a maximum as a function of ω, which is defined by some average rate. According to the above discussion of the temperature dependence, one expects a crossover from the activated behavior at high temperature, (5.125), to a power law at low temperature.

In Figs. 5.13 and 5.14 we show data observed for 60 ppm ^{7}Li and 70 ppm ^{6}Li in KCl. First consider the dielectric loss (which is proportional to the dissipative part of the susceptibility.) For both isotopes, one finds a pronounced relaxation peak at about 100 mK; the peak temperature for the ^{7}Li sample is rather smaller, in quantitative agreement with (5.127) and $^6\Delta_0/^7\Delta_0 = 1.55$. Despite the smaller impurity concentration, the weight of the relaxation peak is significantly larger for ^{7}Li, again in accordance with the theory.

In order to illustrate the above discussion, we evaluate the relaxation spectrum for a simplified distribution of the rates. In (5.123) and (5.124), the prefactors involving f, κ, and I depend in a complicated fashion on the local configuration. Here we resort to a simple approximation by taking

$$\gamma = r\frac{\pi}{\hbar}\left(\frac{f_P^2 \kappa_P^2}{\Delta_0^4} k_B T + \frac{f_A^2 \kappa_A^2}{\Delta_0^3} \cosh(\Delta_0/2k_B T)^{-2}\right) \equiv r\gamma_0, \tag{5.128}$$

with r varying from 0 to 1 with constant probability, $P_r(r) = 1$ for $0 \leq r \leq 1$. When we replace the amplitudes R_i by the mean value $\overline{R}$ and perform the configurational average by integrating over r, we easily find

$$\chi''_{\text{rel}}(\omega) = \overline{R}\frac{np^2}{3\epsilon_0}\frac{1}{k_B T}\frac{\omega}{2\gamma_0}\log\left(1 + \frac{\gamma_0^2}{\omega^2}\right). \tag{5.129}$$

In Fig. 5.12 we plot the relaxation spectrum for 60 ppm Li impurities, and for an external frequency $\omega/2\pi = 10$ kHz. For the unknown factors in (5.128) we take $\sqrt{f_P \kappa_P}/k_B = 27$ mK and $\sqrt{f_A \kappa_A}/k_B = 0.11$ K. Note that these constants are independent of tunnel energy and temperature. (The resulting maximum prefactor (i. e., $r = 1$) defined in (5.125) reads for ^{7}Li as $\Gamma_0/2\pi = 40$ MHz.)

When we compare Figs. 5.12 and 5.13, we find a qualitative agreement with respect to the dependence on tunnel energy and temperature. Note, however, that the actual rate distribution is much more complicated than that of the parameter r in (5.128). Yet the relaxation feature is little affected by the prefactor of the rate γ_A, (5.123); any other smooth parameter distribution would yield similar results. (Compare [62].)

The isotope effect on both relative weight and position of the relaxation peak provides a proof for the Arrhenius law with an activation energy given by the tunnel amplitude Δ_0. Regarding the width of the relaxation peak, if we took the distribution of energies into account (as shown for $\mu = 0.1$ in Fig. 5.5 on p. 85), this would broaden the peaks in Fig. 5.12 and improve the accord with the data. Finally we remark that the theoretical curves of Fig. 5.12 are calculated for a given concentration; if we account for the slight difference in impurity density of the samples, we would obtain values for the relaxation amplitudes close to those deduced from Fig. 5.13.

The constant relaxation spectrum below 50 mK in Fig. 5.12 arises from the linear temperature dependence of the rate γ_P, (5.124). Since in this range the theoretical curves are most sensitive to the distribution of the parameters f, κ, and I, Fig. 5.12 merely indicates that the present approach is suitable for reproducing the observed low-temperature behavior; the actual distribution law is more complicated and involves at least one additional parameter,

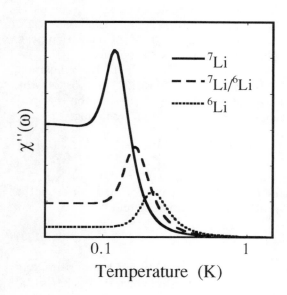

Fig. 5.12. Relaxation spectrum $\chi''_{\text{rel}}(\omega)$ according to (5.129) for 60 ppm Li impurities as a function of temperature. For the parameters of the rate see main text

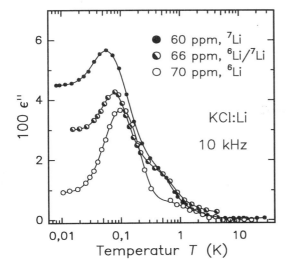

Fig. 5.13. Imaginary part of the dielectric susceptibility for 60 ppm ^{7}Li and 70 ppm ^{6}Li in KCl, and a mixed sample. By courtesy of Weis and Enss [65]

namely the low-energy excitations E_-. At very low temperatures of about 10 mK, the linear law (5.124) ceases to be valid, and the rate γ_P vanishes exponentially. The relaxation feature is little affected by the prefactor of the rate γ_A, (5.123); any other smooth distribution would yield similar results.

For the parameters of the present samples we calculate from (5.126) the ratio $^7\overline{R}/^6\overline{R} \approx 2$. Although it is not an obvious step to relate this number rigorously to the data of Fig. 5.13, there is little doubt that the isotope dependence of (5.126) provides the proper explanation for the observed weights. Finally, we call the reader's attention to the behavior between 10 and 50 mK, where both samples show a weakly temperature-dependent dielectric loss, which is well accounted for by assuming an appropriate distribution for the rate γ_P.

As to the frequency dependence of the maximum temperature T^*, (5.127), it proves to be instructive to compare the data measured at 10 kHz and 650 MHz, as shown in Figs. 5.13 and 5.11. For the smaller frequency, 10 kHz, the relaxation maximum of 60 ppm ^{7}Li occurs at about 60 mK, whereas the 650 MHz data on p. 93 yield $T^* \approx 600$ mK. The difference of a factor of 10 is quantitatively accounted for by (5.127). [Yet the occurrence of a relaxation peak at 650 MHz indicates a smoother distribution of rates than (5.128).]

Now we turn to the real part of the susceptibility. For $\mu \ll 1$, the resonant contribution exceeds by far the relaxational one. Thus χ' is dominated by the resonant contribution, contrary to the imaginary part χ''. In Fig. 5.14 we show data observed for 60 ppm ^{7}Li in KCl at two frequencies, 10 kHz and 650 MHz. First consider the high-frequency data.

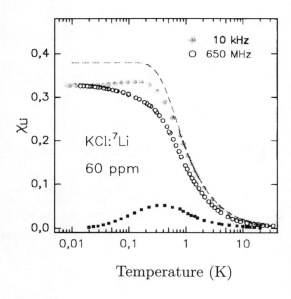

Fig. 5.14. Real part of the dielectric susceptibility for 60 ppm ^{7}Li in KCl at 10 kHz and 650 kHz. The black circles give the difference of both. The dashed line shows the result for non-interacting impurities, χ_{iso}. By courtesy of Weis and Enss [62]

The external frequency $\omega/2\pi = 650$ MHz is significantly larger than most relaxation rates γ, according to (5.109). At sufficiently low temperatures, the relaxational contribution is negligible in this case. The resonant part differs from that of non-interacting impurities in two respects: first, the smeared-out density of states results in a superposition of tanh functions with different energies in the argument, as in (5.98). Second, the reduction of the amplitudes $R_\pm$ [see (5.89)] results in a zero-temperature value, which is smaller than that of isolated defects (dashed line). Both features are clearly seen in Fig. 5.14. Note in particular that the pair model predicts an increase in the susceptibility as compared with non-interacting impurities, in clear contradiction to the measured data.

Regarding the low-frequency data at $\omega/2\pi = 10$ kHz, the real part of the resonant susceptibility (5.94) varies little with frequency; thus χ' looks very much the same for both sets of data. Yet for $\omega/2\pi = 10$ kHz, the relaxation contribution (5.109) gives rise to an additional feature with a maximum at about 300 mK. The temperature dependence of the difference between the two data sets (full squares) confirms the existence of a relaxation feature, as discussed above.

Relaxation at Moderate Concentration $\mu \approx 1$

For $\mu \approx 1$, both the relaxation amplitude R and the finite energy residues $R_\pm$ may be of the order of unity. When calculating the memory function $N''(0)$, the most relevant contribution to (5.59) involves both types of spectra; in terms of the residues it is classified by $R_\pm^{(i)} R_\pm^{(j)} R_k$. Proceeding as for the dilute case, we replace the dipolar coupling term $J_{ij}^2 J_{ik}^2 I_i^{-2}$ by the parameter f^2. With $R_k \approx 1$, we find

$$\hat{N}''(0) = \frac{\pi}{\hbar} \frac{f^2}{\Delta_0^2} \sum_\pm R_\pm E_\pm^2 D(E_\pm) \cosh(E_\pm/2k_\mathrm{B}T)^{-2}. \tag{5.130}$$

For $\mu \approx 1$, the energies $E_\pm$ show a smooth distribution (see Fig. 5.5), and (5.89) reveals that both $R_\pm$ are small; with the prefactor in (5.73), we finally obtain the rate

$$\gamma_\mathrm{A} = \frac{\pi}{\hbar} \frac{f^2 \kappa^2}{\Delta_0^4 + \kappa^2(I^2 + \Delta_0^2)} \sum_\pm R_\pm E_\pm^2 D(E_\pm) \cosh(E_\pm/2k_\mathrm{B}T)^{-2}, \tag{5.131}$$

which is quite similar to that derived above. The main difference concerns the activation energies $E_\pm$, which now cover a large range, from well below Δ_0 to significantly larger values. Note, however, that very small energies (below 100 mK) are missing; the density of states plotted on p. 85 clearly shows the suppression of low-energy pair excitations.

Regarding the isotope effect, since the mean values of both κ and I are close to the tunnel energy Δ_0, Eq. (5.131) does not permit an unambiguous statement. Certainly the prefactor diminishes for larger Δ_0; yet this dependence competes with that of the residue factor $R_\pm$, which for sufficiently large I and κ, is proportional to Δ_0^2 [see (5.89) on p. 87].

Relaxation at High Concentration $\mu \gg 1$

One still finds a contribution (5.131) to the rate. Since R_- vanishes much faster with increasing μ than R_+ does, we retain only the latter term; (5.84) yields $E_+ \approx I$. The rate thus simplifies to

$$\gamma_\mathrm{A} = \frac{\pi}{\hbar} f'^2 R_+ D(E_+) \cosh(E_+/2k_\mathrm{B}T)^{-2}. \tag{5.132}$$

Note that both R_+ and $D(E)$ vary roughly as $\Delta_0^2/I_c^2 = \mu^{-2}$ and thus decrease with rising concentration.

We now briefly discuss temperature dependence and isotope effect. The thermally activated rate γ_A shows a broad distribution of activation energies; accordingly, one expects the relaxation peak to be less sharp than for the dilute case. Since at a given temperature only rates $\gamma \approx \omega$ contribute significantly to relaxation, the spread of the activation energies leads to an apparent reduction in the relaxation amplitude. In order to recover the whole spectrum, one would have to integrate the dissipative part $\chi''_\mathrm{rel}(\omega)$ over many orders of magnitude in frequency.

Contrary to the result of the dilute case, Eq. (5.131) does not depend in an unambiguous way on the isotope mass for $\mu \approx 1$. The prefactor shows a Δ_0^{-4} power law for $\mu \ll 1$ and is independent of Δ_0 for $\mu \gg 1$; in the latter case, both the residues and the coupled density of states are proportional to Δ_0^2.

The average relaxation amplitude $\overline{R}$ shows no isotope effect for $\mu \geq 1$, since it is close to unity in any case. A general statement about the weight of the relaxation peak as a function of the isotope mass would seem impossible because of its intricate dependence on the rate distribution.

All rates derived so far involve a resonant interaction of adjacent tunneling impurities. Yet in the high-density limit $\mu \gg 1$, a different process may prevail; the corresponding rate should be determined self-consistently from the low-frequency correlation spectra of the neighboring impurities. For $\mu \gg 1$, and hence $\kappa \gg \Delta_0$, the relaxation amplitude (5.69) is of the order of unity for most systems,

$$R = \frac{I^2}{\Delta_0^2 + I^2}; \tag{5.133}$$

accordingly, the finite-energy residues $R_\pm$ are small, and so is the coupled density of states $D(E)$. Thus one expects a more significant contribution to (5.59) arising from the relaxation parts of all three correlation spectra $G'''(\omega)$,

$$\hat{N}''(0) = 2 \sum_{j,k \neq j} \frac{J_{ij}^2 J_{ik}^2}{\Delta_0^2 I_i^2} R_i R_j R_k \left(\frac{\omega^2 \gamma_i}{\omega^2 + \gamma_i^2} * \frac{\gamma_j}{\omega^2 + \gamma_j^2} * \frac{\gamma_k}{\omega^2 + \gamma_k^2} \right). \tag{5.134}$$

The rates γ are much smaller than the thermal frequency $k_B T/\hbar$; hence the weight factors in the convolution integrals (A13) are immaterial. Approximating the prefactor by Δ_0^2/I^2, we find

$$\gamma_i = 2\pi \sum_{j,k \neq j} \frac{J_{ij}^2 J_{ik}^2}{I_i^4} \frac{\gamma_i^2}{\gamma_i + \gamma_j + \gamma_k}. \tag{5.135}$$

Both the factor R_+ and the coupled density of states in (5.131) are small in the limit $\mu \gg 1$; thus for such high concentrations the rate should be determined by (5.134).

Equation (5.135) constitutes a set of self-consistency relations for the rates; these are a simplistic version of the integral equations resulting quite generally from a mode-coupling approach to the relaxational dynamics of disordered systems. Assuming all rates are the same, (5.135) reduces to $1 = (2\pi/3) \sum'_{j,k} J_{ij}^2 J_{ik}^2 I_i^{-4}$, which is obviously inappropriate for determining the relaxation rate. As a second flaw, we note that (5.135) does not depend on temperature.

Any attempt to derive the relaxation dynamics for the case $\mu \gg 1$ would require a more detailed investigation of the mode-coupling equations (5.21) and (5.59). Similar but not identical integral equations have been derived by Götze in his treatment of the spin glass and glass transitions [122,178].

As one essential difference, we note the occurrence of a factor ω^2/Δ_0^2 in the convolution integrals of the memory functions (5.58) and (5.59); this may be traced back to the fact that the interaction term in the Hamiltonian (5.1) involves only one spin component, σ_z, and that the time derivative of σ_z does not involve the dipolar coupling energy J_{ij}. This factor ω^2/Δ_0^2 shows up in (5.134) and leads to the factor γ_i^2 on the right-hand side of (5.135), which renders it unsuitable for determining the relaxation rate; without that factor, the self-consistency relation would yield a quadratic equation for γ. The spin glass and glass models exhibit a coupling of all spin components,

$$J_{ij}\left(\sigma_x^i\sigma_x^j + \sigma_y^i\sigma_y^j + \sigma_z^i\sigma_z^j\right), \tag{5.136}$$

and thus do not encounter that problem.

Since it involves all three spin components, the interaction of the spin glass model gives rise to non-trivial dynamics. On the contrary, the interaction term in (5.1) has no dynamics at all. Only the tunneling part permits motion of the impurities.

The presence of the second energy scale Δ_0 in our Hamiltonian (5.1) is essential. Most of our results involve a cross-over arising from the competition between tunnel and coupling energies; their ratio defines the parameter μ. The spin glass and glass models deal with a non-ergodicity transition which occurs where the coupling energy exceeds temperature.

In summary, we expect the nature of the present problem to change qualitatively for $\mu \gg 1$. In that case, the tunnel energy would seem to be rather irrelevant; this is supported by the fact that Δ_0 has disappeared from (5.135).

5.9 Discussion

Projection Scheme

The mode-coupling approximation relies on the representation of the resolvent matrix element (5.12) as a continued fraction. The number of projections determines the number of poles in the correlation function. Thus performing four projections led to the polynomial of the fourth order in the denominator of (5.67).

The operators appearing in the continued fraction, $\sigma_z^i, \dot{\sigma}_z^i, \ddot{\sigma}_z^i, ...$ involve composite operators of any order. It proves to be essential to treat the contributions containing one or two spins separately; they are gathered in the set of relevant operators $A_1,...,A_4$. Accordingly, the dynamics of a coupled pair of defects is described exactly by restricting the continued fraction to these two-mode contributions. Composite operators of higher order are taken into account by the memory functions. (Note that both N and M involve two different neighbors, $j \neq k$.)

The memory functions affect the dynamical behaviour in two different ways. First, $M(z)$ contains a part $K(z)$ which is singular at $z \approx 0$, resulting in a relaxation pole in $G(z)$. Second, the remaining contributions $\hat{M}(z)$ and $N(z)$ add imaginary parts to the poles of $G(z)$, thus leading to exponential damping of the correlation function $G(t)$.

Approximations

The mean relaxation amplitude $\overline{R}$ is a most significant quantity; it has been evaluated in both three-pole and five-pole approximation by either performing the average of I or setting up a self-consistency equation (Appendix D). Any of these approaches yields a cross-over from coherent oscillations to relaxation dynamics at $\mu \approx 1$.

Mori theory and resulting self-consistent mode-coupling equations have been studied in detail [177–179]; the emergence of a zero-frequency peak was first derived by Götze [177] and discussed extensively in [178]. The reduction of the spectral weight of the finite resonance energies in favor of the relaxation peak is supported by general considerations given in these works.

A priori it is not obvious that the spectrum should comprise an odd number of poles. Yet if we truncated the continued fraction (5.18) at second or fourth order, we would obtain unphysical results such as, e.g., too large damping functions. (At low temperatures, relaxation and damping rates must be small, however, even if the motion is incoherent.) Comparing the results from the schemes involving three and five poles, we find them to agree regarding the essential features, as shown in Figs. 5.2 (p. 72) and 5.15 (p. 111).

The three-pole approximation reduces in the low-density limit to the dynamics of a two-level system; accordingly, it does not account for low-energy excitations of nearby impurity pairs. Pushing the continued fraction to higher order yields a five-pole spectrum for the correlation function; as a major advantage the latter scheme permits us to recover the pair model at low concentrations. At high impurity density, or $\mu \gg 1$, both approximations yield purely relaxational dynamics.

Obtaining explicit expressions for the spectra requires some additional approximations; besides the decoupling of static expectation values in (5.45, 5.46), we note the average over the quantity κ^2 and the mean-field treatment with respect to the random defect configuration. In particular, the latter scheme is not unique. In [60] the quantity κ has been treated in a different approximation, leading to almost identical results, except for the low-energy density of states. (Yet the resulting correlation of κ and I constitutes an unsatisfying feature of [60].) As another alternative to the most simple approximation (5.60), we have set up a self-consistent equation for the average relaxation amplitude $\overline{R}$ (see Appendix D); again, we found results essentially identical to those obtained from (5.60).

Cross-Over at $\mu \approx 1$

Very early it was realized that increasing the impurity density beyond a few hundred ppm led to a significant change in the thermal behavior. At higher concentration the picture of independent defects failed to explain the observed phenomena, hinting at the crucial role of interaction [54,37]. In analogy to the critical temperature of a ferro-electric or ferro-elastic phase transition, a cross-over temperature was defined through the average interaction by taking $k_B T_c = (np^2/4\pi\epsilon\epsilon_0)$. This approach indeed accounted for the concentration dependence of the maximum temperature of the relaxation peak, which was observed at high concentration. Yet no ordering of the impurity dipoles was found, and the susceptibility did not diverge as expected at a phase transition; quite on the contrary, it decreased when the impurity concentration was augmented.

The present approach does not yield a critical temperature, but rather reveals a smooth change as a function of defect density. Both three-pole and five-pole approximations yield a cross-over to relaxational dynamics at an impurity density n where the average interaction energy

$$\frac{2}{3}\frac{np^2}{\epsilon_0\epsilon} \equiv I_c \qquad (5.137)$$

is roughly equal to the tunnel energy Δ_0, corresponding to $\mu \approx 1$ in terms of the coupling parameter μ. Clearly, the limiting cases of small and large μ show a basically different temperature dependence of thermodynamic quantities, but this is not related to a phase transformation.

In the limit $\mu \gg 1$ the dynamics is entirely determined by relaxation. The two approximation schemes give the same amplitude, (5.23) and (5.133), with an average value tending towards unity, according to (5.27) and (5.113). The rates derived from 5-pole approximation agree with previous results based on a 3-pole correlation spectrum [59].

When we consider the dilute limit $\mu \ll 1$, the 3-pole scheme reduces to the well-known susceptibility of a set of independent two-level systems, as is seen in (5.29) for the limit $R_i \to 0$. From the 5-pole approximation, however, one recovers the pair model with two different resonances $\eta_\pm$. Obviously, at very low concentrations the latter, in turn, factorizes in a set of N identical two-level systems. Yet in the intermediate range, $\frac{1}{10} \leq \mu \leq 1$, there are significant deviations from the two-level picture; thus constructing a theory with the pair model as the low-density limit would not seem an idle undertaking. Considering that most experiments are done on samples with impurity concentrations corresponding to $\frac{1}{10} \leq \mu \leq 1$ confirms the relevance of this range.

The cross-over is driven by the quantity I^2 in three-pole and by κ^2 in five-pole approximation. From (5.47, 5.63), it is clear that both are determined by the average interaction energy (5.137). The expressions for the amplitudes (5.23, 5.133) and for the energies (5.22, 5.85) closely resemble the well-known results for a two-level system with finite asymmetry energy. (Compare (6.6) and (6.9), respectively.) As an essential difference, we emphasize that in the present case the static correlation $\langle\sigma_z\rangle$ vanishes; thus the quantity κ, or I, may be considered as a dynamic asymmetry.

In the remainder of this section, we summarize the main features of density of states, specific heat, and dielectric susceptibility.

5. Cross-Over to Relaxation: Mori Theory

Density of states. In the zero-density limit there is a single excitation energy Δ_0. At small but finite concentrations, most impurities are weakly coupled and thus show resonance energies close to the bare tunnel splitting Δ_0; accordingly, the density of states $D(E)$ shows a double-peak structure at $\Delta_0 \pm \frac{1}{2}I_c$. A few impurities, however, form nearby pairs resulting in the tails at low and high energies. Both resonances $E_\pm$ and residues $R_\pm$ are well approximated by those derived in the previous chapter for the pair model, (4.22, 4.54). In general the pair model provides a good description for the dilute limit. Note, however, that it misses the relaxation peak.

As compared to the density of states of the pair model, at moderate concentration $\mu < 1$, the quantity κ introduces two novel features. First, it reduces the weight of the low-energy tail of $\rho(E)$, as is displayed by the full curve in Fig. 5.5, and then it gives rise to a finite relaxation amplitude, (5.69).

As μ exceeds unity, the fate of the low-energy tail is the most important aspect of the present approach. The pair model exhibits an ever-increasing spectral weight at low energy, whereas in the Mori approach the corresponding weight of the coupled density of states is transferred to an incoherent contribution at zero frequency.

For $\mu \gg 1$, the relaxation amplitude of almost all impurities is close to unity, and the finite-energy residues $R_\pm$ are negligibly small. Both energies $E_\pm$ are significantly larger than the bare tunnel energy Δ_0. Whereas the larger energy is given by the interaction of the nearest neighbor, $E_+ \approx I$, the smaller one is determined by the next-nearest-neighbor energy, $E_- \approx \kappa$.

Specific heat. The specific heat as a function of temperature reflects all features of the density of states, as is most evident when comparing Figs. 5.5 (p. 85) and 5.7 (p. 88). At low density, or $\mu \ll \frac{1}{10}$, there is a single excitation energy Δ_0, leading to the narrow Schottky anomaly. With rising impurity concentrations, one observes an excess specific heat at very low temperatures, and a significant broadening of the Schottky peak. Figure 5.10 shows that the data measured for concentrations above 200 ppm rule out the density of states derived from the pair model. On the other hand, Mori theory with a mode-coupling approximation provides an almost quantitative explanation.

Dynamic susceptibility. The most striking experimental observation involves the static susceptibility. For very low densities, or $\mu \ll 1$, one finds a linear increase with the number of impurities. With rising concentrations the dipolar interaction becomes significant and leads to a weaker increase and even, above the value $\mu \approx 1$, to a decrease of the susceptibility. (See Fig. 5.1.) The Mori approach accounts well for this vanishing of the susceptibility at $T = 0$ and $\omega = 0$, as shown in Fig. 5.3 on p. 74 for various impurity systems.

The static susceptibility is closely related to the coupled density of states. At small concentrations, the Mori approach yields a density of states similar to that obtained from the pair model. As μ approaches unity, however, the low-energy spectral weight of the coupled density of states is transferred to

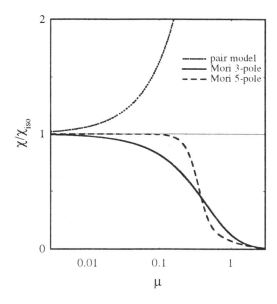

Fig. 5.15. Static susceptibility at zero temperature as a function of the coupling parameter μ, as obtained from 3-pole and 5-pole approximation of Mori theory by (5.32) and (5.106), for the pair model by (4.70), and for non-interacting impurities. The solid line is the same as in Fig. 5.3.

an incoherent contribution at zero frequency, resulting in the relaxation peak of the susceptibility and in a decrease of the static susceptibility.

In Fig. 5.15 we plot the latter quantity as obtained from different approximations for the Mori projection scheme and from the pair model. As in Fig. 5.3, the susceptibility is normalized with respect to its value for isolated defects, (4.11). In the low-density limit $\mu \ll 1$, all schemes reduce to the simple expression for vanishing interaction, χ_{iso}, whereas they lead to different results for high density, or $\mu \geq 1$. The normalized susceptibility χ/χ_{iso}, as derived from the pair model, steadily *increases* with impurity concentration. On the contrary, the Mori approach predicts a strong *decrease* with rising impurity density, in agreement with experiment.

Another sign for the cross-over is provided by the average relaxation amplitude $\overline{R}$. For $\mu \ll 1$ it is much smaller than unity, at about $\mu \approx 1$ it strongly increases and finally tends towards unity. For (5.28) one finds the integrated relaxational absorption

$$\int d\omega \chi''_{\text{rel}}(\omega) = \overline{R}\frac{2\pi}{3}\frac{np^2}{\epsilon_0 k_B T} \tag{5.138}$$

to increase with the inverse of the temperature. This law basically differs from the absorption of a two-level system with asymmetry Δ and energy splitting $E = \sqrt{\Delta_0^2 + \Delta^2}$,

$$\int d\omega \chi''_{\text{rel}}(\omega) = \frac{\Delta^2}{E^2}\frac{2\pi}{3}\frac{np^2}{\epsilon_0 k_B T}\frac{1}{\cosh(E/2k_B T)^2} \quad \text{for TLS,} \tag{5.139}$$

which vanishes exponentially for $k_B T \ll E$.

Weis and Enss have measured the dielectric loss of KCl:Li from 100 Hz to 100 kHz at various temperatures between 9 mK and 300 mK. Over this frequency range, the absorption does not vary significantly, and it does not depend strongly on temperature. Since the spread of relaxation rates exceeds the limited frequency range, the available data merely provide a lower bound for the relaxation amplitude. It would seem, however, that the data do not agree with an exponential law as in (5.139), but rather show a smoother temperature dependence as in (5.138).

In summary, we find the following picture for the dynamics of a tunneling defect as a function of the parameter μ. For small concentrations, or $\mu \ll 1$, the correlation function $G(t)$ shows oscillations with frequencies $E_\pm/\hbar$. With rising concentrations, the lower branch E_- shifts to larger energies and leads to an increasing average energy; a relaxation peak emerges at $\omega \approx 0$. For $\mu > 1$ we find that the spectral weight of the oscillations is suppressed and the dynamics are governed by incoherent tunneling.

Comparison with the pair model. For low density $\mu \ll 1$, the Mori approach reduces to the pair model presented in Chap. 4; at the opposite limit they are basically different. The continued fraction of Mori theory gives rise to a relaxation pole that absorbs the spectral weight of the low-energy excitations, thus reducing significantly the dielectric susceptibility as compared to the value for non-interacting impurities.

The increasing weight of low-energy excitations of the pair model results in a zero-frequency susceptibility, which, for large μ, exceeds that of isolated impurities by a factor of μ, i.e., the susceptibility increases with the square of the impurity density. The limiting expression (4.72)) is clearly unphysical; it is invalidated by experiment.

Classical Limit at High Temperatures

At sufficiently high temperatures, quantum effects are immaterial. When we evaluate (5.93, 5.98, 5.109) for $E_\pm \ll k_B T$ and at low frequency $\omega \ll \gamma$, we find that the dielectric susceptibility derived in this chapter reduces to the classical expression

$$\chi_{\text{class}}(\omega \to 0) = \frac{1}{3} \frac{np^2}{4\pi\epsilon_0} \frac{1}{k_B T}, \qquad (5.140)$$

independent of the value of μ.

5.10 Summary and Outlook

In this chapter we have investigated how interaction affects the tunneling dynamics when the impurity concentration is augmented. Here we briefly summarize the main results:

(a) Existence of low-energy excitations. The resonant absorption data shown in Fig. 5.11 prove the existence of low-energy pair excitations; temperature dependence, prefactor, and isotope effect are in accordance with (5.97), thus confirming the validity of our model.

(b) Cross-over at $\mu = 1$. For the mean interaction energy $(2np^2/3\epsilon\epsilon_0)$ approaching the tunnel amplitude Δ_0, we find a cross-over to relaxational dynamics. As its most striking feature, we note the reduction of the dielectric susceptibility as shown in Fig. 5.3 on p. 74, and the broadening of the Schottky anomaly of the specific heat in Figs. 5.9 and 5.10.

(c) Relaxation driven by dipolar interaction. At a lithium concentration as small as 60 ppm, dynamic measurements exhibit a relaxation contribution. The dependence on both temperature and isotope mass rule out phonon scattering from asymmetric two-state systems as a relaxation mechanism, but rather indicate the relevance of the dipolar interaction.

(d) Isotope effect for the relaxation rate. The observed relaxation rate for the heavier isotope ^{7}Li is about five times larger than that for ^{6}Li. This unusual isotope effect is explained by the model of interacting impurities. (For phonon scattering one would expect the inverse dependence [53].)

There remain a number of open questions, mainly concerning the relaxational dynamics at higher concentrations. In the previous section Mori theory has been shown to account, at least qualitatively, for the observed temperature dependence. Yet a thorough comparison would require us to investigate the rate distribution in detail.

Moreover, we have not addressed the resulting frequency dependence of the susceptibility. Experimentally, one finds a stretched exponential behavior, with an additional low-frequency tail arising from slowly relaxing systems [65]. Again, our simplistic treatment leading to (5.129) accounts for the main features of the observed spectra. Yet one would like to know whether this agreement could be made more quantitative.

Finally, as discussed on page 106, for $\mu \gg 1$ the present approach gives rise to a self-consistency problem similar to Götze's spin glass model. Further work would seem interesting in two respects. First, it is not clear whether the different interaction term in the Hamiltonian (5.1) results in the same dynamical behaviour as a well-known spin glass model. Second, deriving the relaxation spectrum from the mode-coupling equations would permit a thorough comparison with measured data and to settle the question concerning a non-ergodic low-temperature phase.

6. The Tunneling Model

6.1 Low-Temperature Properties of Glasses

At low temperatures the thermal, acoustic and dielectric behavior of glasses differs significantly from that of crystalline solids. As the most prominent examples, we note the almost linear temperature dependence of the specific heat [106] and the surprising variation of sound velocity with frequency and temperature [107].

Well below 1 K the specific heat of most insulating glasses exceeds that of dielectric crystals by orders of magnitude, and over several decades it obeys a power law as a function of temperature with an exponent close to unity. Figure 6.1 shows data for vitreous silica with different OH concentrations; the

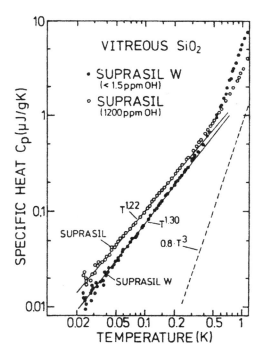

Fig. 6.1. Specific heat for vitreous silica below 1 K. The Debye contribution is indicated by the dashed line. By courtesy of Lasjaunias et al. [108]

6. The Tunneling Model

dashed line indicates the Debye contribution to the specific heat arising from acoustic lattice vibrations. On the other hand glasses are bad heat conductors; the thermal conductivity is smaller than for crystals and shows a power law behavior with an exponent close to two, whereas for pure crystals one finds a value of three [106,109].

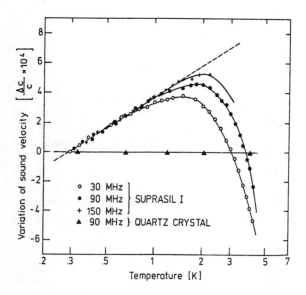

Fig. 6.2. Relative variation with temperature of the longitudinal sound velocity $\Delta c/c$ for vitreous silica and quartz below 1 K. The solid lines are fits according to the perturbation theory presented in the next chapter; the dashed line gives only the logarithmic increase due to resonant interaction. By courtesy of Piché et al. [110]

The difference between amorphous and crystalline insulators is even more pronounced when considering dynamic quantities. Figure 6.2 displays the longitudinal sound velocity as a function of temperature measured at several frequencies. For a crystalline solid, thermal expansion leads to a weak decrease with rising temperatures; the corresponding variation for quartz crystal is imperceptible on the scale of Fig. 6.2. The change in sound velocity observed for vitreous silica is larger by many orders of magnitude; below 1 K one finds a logarithmic increase with temperature; then the slope changes sign, and at higher temperatures the sound velocity strongly decreases.

Acoustic waves in crystals travel at the same velocity, independent of frequency. The data of Fig. 6.2 reveal that glasses behave differently. Regarding the logarithmic increase below about 1 K, the data for 30, 90, and 120 MHz coincide. Yet both the maximum temperature and the subsequent decrease depend on the applied frequency. For frequencies in the kHz range, the maximum occurs at a temperature of about 100 mK; above this temperature, the change of sound velocity obeys a logarithmic law with negative slope up to about 1 K [128].

6.1 Low-Temperature Properties of Glasses   117

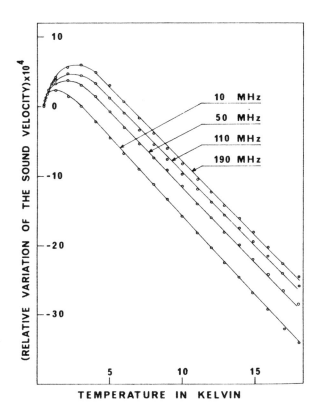

Fig. 6.3. Relative variation of the longitudinal sound velocity $\Delta v/v$ with temperature for vitreous silica between 0 and 20 K. The absolute value is $v = 5.8 \times 10^5$ cm/s. From [136]

A qualitative change occurs at about 5 K. Above this temperature, the sound velocity decreases linearly, as shown for vitreous silica in Fig. 6.3. Although this behavior has been observed unambiguously for various materials, it has been paid little attention in the literature on low-temperature properties of glasses. As further examples we note amorphous selenium [137] and amorphous metallic NiP [138]. For data on oxide glasses see, e.g., [110,128]; there a flattening of the linear law has been found for temperatures above 25 K. As the second most remarkable feature, the slope of the sound velocity with respect to temperature depends nearly logarithmically on frequency.

Sound attenuation (or 'internal friction') exhibits a characteristic temperature dependence as shown schematically in Fig. 6.4, whose main features are closely related to those of the sound velocity. At very low temperatures one finds a power law increase, which saturates to a constant above a temperature $\widetilde{T} \approx 200$ mK, corresponding to the maximum of sound velocity displayed in Figs. 6.2 and 6.3. After a plateau region of about one decade in temperature the internal friction starts to increase again at a temperature

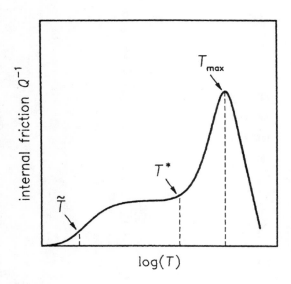

Fig. 6.4. Schematic temperature dependence of the internal friction in amorphous solids. By courtesy of Rau et al. [128]

$T^* \approx 5$ K, which, in turn, is related to the onset of the linear decrease in sound velocity. At still higher temperatures, it reaches a relaxation maximum of $T_{\max} \approx 30$ K, and then falls off rapidly. As typical examples we discuss in the next chapter oxide glasses and polystyrene (see Figs. 7.1 and 7.6).

6.2 The Tunneling Model

Various substances materialize both in a crystalline and in an amorphous phase; as examples we note oxides of silicon, germanium, and boron, and several metals. (Accordingly, one has to distinguish between insulating and metallic glasses.) Most insulating amorphous materials show similar low-temperature properties, which differ in essential aspects from those of the crystalline counterparts; an illustrative example is provided by Fig. 6.2. This indicates that the low-temperature anomalies are not due to some peculiar chemical or physical feature of a given compound, but rather originate from the configurational disorder. The difference between the crystalline and amorphous states of silica is shown schematically in Fig. 6.5.

Various attempts have been made to explain the low-temperature anomalies of glasses; dangling atoms in small accidental cages [100], 'frozen-in' local density fluctuations [101], and strongly damped low-frequency vibrational modes [102] have been put forward. None of these approaches, however, could provide a clear physical picture accounting for the observed anomalies.

The most successful model up to now has been proposed independently by Anderson, Halperin, and Varma [103] and by Philips [104]. In order to ex-

6.2 The Tunneling Model

plain the unusual temperature dependence of the specific heat, these authors postulated localized low-energy excitations arising from quantum tunneling between metastable atomic configurations. Soon afterwards Jäckle derived a relaxation contribution from these tunneling systems and thus resolved the anomalous temperature variation of the sound velocity [105,110] (cf. Fig. 6.2). This tunneling model for glasses relies on two assumptions, which are plausible and both have been confirmed by many experiments since; they nevertheless lack a microscopic derivation.

First, the potential energy of the atoms is thought to exhibit almost degenerate minima at sufficiently small distances and when separated by a sufficiently low barrier, in order to permit the system to move by quantum tunneling from one configuration to the other. In the simplest case one has a double-well potential as shown schematically in Fig. 6.6.

Second, the parameters of such a potential obey a particular distribution, which, moreover, is thought to be the same for various glasses, irrespective of their precise microscopic structure.

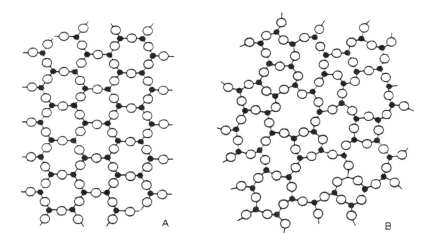

Fig. 6.5. Ordered and disordered state for SiO_2, schematically. From [113]

Regarding the first assumption, Fig. 6.5 suggests the occurrence of chemical bonds that have spacings and angles at far from the most favorable values in terms of energy. For a sufficiently large number of such frustrated bonds, the existence of almost degenerate distinct configurational states seems natural; the stiffness of directed chemical binding results in a potential barrier between such configurations. The lack of microscopic specificity has been considered as a flaw in the tunneling model; in view of the different materials

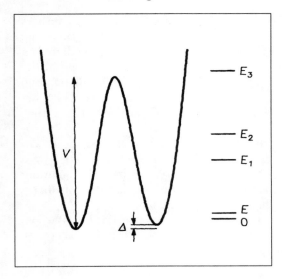

Fig. 6.6. Double-well potential with finite asymmetry energy Δ; in addition to the ground-state splitting, the next few energy levels are indicated. From [128]

exhibiting glass-like behavior, a too specific model might well describe a given compound, but would fail to account for the universal aspects of amorphous solids.

The potential in Fig. 6.6 is a function of some collective coordinate involving displacements of several (or many) atoms; transformation from one low-energy configuration to another will require the reorientation of the local structure. At low temperatures only the ground states in the two wells, $|R\rangle$ and $|L\rangle$, are relevant. In WKB approximation the tunneling amplitude $\Delta_0 \equiv \langle R|V|L\rangle$ between these states is given by

$$\Delta_0 = \hbar\omega_0 e^{-\lambda}, \tag{6.1}$$

where ω_0 is of the order of the Debye frequency, and where the tunnel parameter

$$\lambda = \sqrt{2mVd^2}/\hbar \tag{6.2}$$

depends on the effective mass m, the potential barrier V, and the distance of the wells d. Because of the lack of crystal symmetry, these minima are not degenerate in general; the energy difference between the states localized in the left and the right well is denoted by Δ. For some glasses there are microscopic models for the tunneling systems; thus for amorphous silica the irregular network of Si–O bonds is thought to give rise to metastable states [107].

The operators constructed from the quantum states $|R\rangle$ and $|L\rangle$ are most conveniently expressed through Pauli matrices,

$$\sigma_z \equiv |L\rangle\langle L| - |R\rangle\langle R|, \tag{6.3}$$

$$\sigma_x \equiv |R\rangle\langle L| + |L\rangle\langle R|, \tag{6.4}$$

where the discrete coordinate σ_z takes the values $\sigma_z = \pm 1$. The Hamiltonian thus reads

$$H_0 = -\tfrac{1}{2}\Delta_0 \sigma_x - \tfrac{1}{2}\Delta \sigma_z; \tag{6.5}$$

its eigenvalues $\pm\tfrac{1}{2}E$ are given by

$$E = \sqrt{\Delta_0^2 + \Delta^2}, \tag{6.6}$$

and the statistical operator reads

$$\rho = \frac{1}{2}\left[1 - \tanh(\beta E/2)\left(\frac{\Delta_0}{E}\sigma_x + \frac{\Delta}{E}\sigma_z\right)\right]. \tag{6.7}$$

(We have adopted the common notation for tunneling in glasses; in the theoretical literature on the spin boson model, the tunneling amplitude Δ_0 is usually denoted by Δ, and our asymmetry or bias energy Δ by ϵ. Note, moreover, that in many papers the role of σ_x and σ_z is exchanged.)

Dynamic quantities are described by the two-time correlation function

$$G(t - t') = \tfrac{1}{2}\langle \sigma_z(t)\sigma_z(t') + \sigma_z(t')\sigma_z(t)\rangle; \tag{6.8}$$

with (6.5) one easily calculates

$$G(t) = \frac{\Delta^2}{E^2} + \frac{\Delta_0^2}{E}\cos(Et/\hbar). \tag{6.9}$$

Note that (6.9) exhibits a static part in addition to that oscillating with frequency $E/\hbar$.

For $\Delta = 0$ the particle oscillates between the two wells of Fig. 6.6; if it is initially localized in the left well, it will be found with probability 1 in the right well after a time $\pi\hbar/E$, and after a time $2\pi\hbar/E$ it is back in the left well. For the symmetric case, the eigenstates of the Hamiltonian

$$|\pm\rangle \equiv \frac{1}{\sqrt{2}}(|L\rangle \pm |R\rangle), \tag{6.10}$$

are states of definite parity.

For finite asymmetry $\Delta \neq 0$ the oscillatory motion is incomplete; only the fractional probability Δ_0^2/E^2 is transferred from one well to the other.

6.3 Distribution of the Parameters Δ_0 and Δ

When we start from a constant distribution for asymmetry energy Δ and tunneling parameter λ,

$$P(\Delta, \lambda)\mathrm{d}\Delta\mathrm{d}\lambda = P_0 \mathrm{d}\Delta\mathrm{d}\lambda, \tag{6.11}$$

and replace the dependence on λ by means of (6.1), we easily find [103,104]

$$P(\Delta, \Delta_0) d\Delta d\Delta_0 = \frac{P_0}{\Delta_0} d\Delta d\Delta_0. \tag{6.12}$$

Although introduced in an ad hoc fashion, (6.11) is supported by physical arguments. The asymmetry Δ arises from a random strain field due to local configurational disorder; its distribution is an even function and is roughly constant up to values for Δ of the order of interatomic potentials; thus for small asymmetry it may be replaced by its value at $\Delta = 0$. Regarding the parameter λ, it seems reasonable to assume a smooth distribution for the product of width and height of the double-well potential and for the effective mass; it is easy to show that even a power law $P(\Delta, \lambda) \propto \lambda^x$ in (6.11) would merely cause logarithmic corrections in (6.12), $P(\Delta, \Delta_0) \propto \log(\hbar\omega_0/\Delta_0)^x/\Delta_0$.

It proves to be convenient to change once more to new variables E and

$$r \equiv \Delta_0^2/E^2; \tag{6.13}$$

the resulting distribution

$$P(E, r) dE dr = \frac{P_0}{2r\sqrt{1-r}} dE dr \tag{6.14}$$

renders obvious the constant density of states. (In order to account for the observed deviation of the specific heat exponent $1+\delta$ from unity, with $\delta \approx 0.2$, appropriate modifications of the distributions (6.11–6.14) have been used.) We have not specified the bounds of those distributions. The logarithmic divergence of the integral over Δ_0, or over r, requires a finite lower cut-off; together with the upper bound for E or Δ, this leads to quite involved expressions for the above transformations of variables. Equation (6.14) should rather be considered as a phenomenological law accounting well for experimental observations.

6.4 The Heat Bath

A two-state coordinate σ_z will interact with collective degrees of freedom of the solid. It has been argued that most cases are well described by linear coupling to a set of harmonic oscillators

$$f = \sum_k \lambda_k (b_k + b_k^\dagger), \tag{6.15}$$

where λ_k is a coupling constant and k labels the bath modes [15]. With (6.5) we obtain the Hamiltonian

$$H = H_0 + \frac{1}{2} f \sigma_z + H_\mathrm{B}, \tag{6.16}$$

where $H_\mathrm{B} = \sum_k \hbar\omega_k b_k^\dagger b_k$ is the unperturbed energy of the bath which has operators that fulfil $[b_k, b_{k'}^\dagger] = \delta_{k,k'}$.

Sufficiently dilute tunneling centers hardly affect the bath modes. (The coupling constants vanish as $\lambda_k \propto 1/\sqrt{N}$ with the number of atoms in the sample.) Hence time evolution of the bath operators is determined by H_B only. Accordingly, the heat bath is entirely characterized by the spectral function

$$J(\omega) = \frac{1}{4} \int_{-\infty}^{\infty} dt\, e^{i\omega t} \langle [f(t), f] \rangle = \frac{\pi}{2} \sum_k \lambda_k^2 \delta(\omega - \omega_k). \tag{6.17}$$

Equations (6.16, 6.17) state the so-called spin-boson problem, which, with an appropriate choice for the spectral function $J(\omega)$, accounts for various situations in solid-state physics and chemistry [7,15,23]. Two particularly interesting cases are defined by the linear and cubic spectral functions. (We discuss atomic tunneling only, where the energy Δ_0 never exceeds a few Kelvin. At such low frequencies the spectrum $J(\omega)$ shows a simple power law behavior [15].)

The linear case $J(\omega) = \pi \alpha \omega$ leads to a frequency-independent damping function at low frequencies and finite temperatures; for this reason it is usually referred to as the case of Ohmic dissipation. As the most prominent example we note electron-hole excitations from the Fermi surface; both interstitial hydrogen in metals and tunneling systems in metallic glasses have been described successfully by the Ohmic damping model. As its most striking feature, a logarithmic infrared singularity arises in any order of perturbation theory from the linear frequency dependence. Summing up these logarithmic terms, Kondo derived a characteristic power law for the tunnel energy as a function of temperature, $\tilde{\Delta}_0 \propto T^\alpha$ [6].

Following Langer's path-integral approach to nucleation [8], Caldeira and Leggett derived an exponential damping for the escape of a particle from a metastable state [4]. After Chakravarty noticed [9] a close analogy to the Kondo problem [10], much work was devoted to the dissipative two-state dynamics arising from an Ohmic heat bath, using mainly path integral methods (e.g., [11–23]) and mode-coupling theory [24,25]. The infrared singularity was found to drive a localization transition at $T = 0$. At higher temperatures there is a cross-over from damped oscillations to overdamped motion, with a relaxation rate that, in the weak-coupling regime, decreases with rising temperature [12–16]. Including a cubic part in the spectral function, Grabert derived, at still higher temperatures, a second cross-over to an exponentially increasing rate [23,91].

Screening by conduction electrons has been observed both by acoustic experiments on two-level systems in metallic glasses [26,133] and by neutron scattering on interstitial hydrogen [27].

Now we turn to insulating materials where low-frequency lattice vibrations provide the most efficient damping mechanism. The sound velocity and attenuation data shown in Figs. 6.2–6.4 indicate relaxational motion of some localized degree of freedom. In order to account for the observed relaxation phenomena, one needs to consider the coupling of the two-level systems to the thermal motion of the atoms constituting the glass. In linear order, the coupling energy is given by the elastic strain ϵ and the deformation potential g,

$$g\epsilon = \sum_k \lambda_k (b_k + b_k^\dagger) = -2gk\sqrt{\hbar/2m_k\omega_k}(b_k + b_k^\dagger). \tag{6.18}$$

The two-state coordinate σ_z involves the configuration of many atoms; each two-level system represents a different microscopic arrangement. Hence the coupling term (6.18) is obtained in a less rigorous way than that for a substitutional defect, (2.38); for this reason neither polarization nor propagation directions of the sound waves are indicated. Nevertheless it constitutes the lowest-order term of an expansion regarding the elastic strain arising from sound waves [112].

For acoustic sound waves one finds with (2.36, 2.37) a cubic frequency dependence,

$$J(\omega) = \pi(\alpha/\tilde{\omega}^2)\omega^3. \tag{6.19}$$

A heat bath with cubic spectral density leads to damping phenomena that are basically different from Ohmic dissipation. Whereas the latter case is determined by an infrared singularity, the cubic spectral function does not cause any anomaly at low frequencies; yet regarding the weight of thermal phonons, the increasing $J(\omega)$ causes a strong enhancement with temperature.

Making use of the early work of Holstein on polaron motion [86], Flynn and Stoneham investigated the diffusive motion of a light interstitial in metals [87]; at sufficiently high temperatures, the diffusion constant is determined by a Debye-Waller factor, accounting for thermal lattice vibrations. Kagan and Maksimov started from a delocalized coherent state for the interstitial and studied the destruction of coherence through the lattice motion. The dynamics of a dissipative two-state system were tackled by perturbation theory [88], path integral methods [15,23,91], and mode-coupling theory [89,94–96]. There seems to be general agreement on behavior at low temperatures, where perturbation theory yields a simple picture of weakly damped tunneling oscillations; in the case higher temperatures, contradictory results have been obtained through different approaches.

We deal with tunneling systems in insulating glasses and hence address only the case of a cubic spectral function that describes coupling to sound waves. Concerning Ohmic damping, we refer to review articles by Leggett et al. [15] and, for hydrogen in metals, by Grabert and Schober [23], and for technical aspects to the book by Weiss [21].

7. Tunneling Systems in Amorphous Solids

In this chapter we treat the dissipative dynamics of a two-level tunneling system in the limits of both weak and strong coupling by means of Mori's projection method. In the weak-coupling case, we obtain results well-known from perturbation theory. When we resort to a mode-coupling approximation, we find a qualitatively different behavior in the case of strong coupling or high temperatures. Then we apply our theory on tunneling systems in glasses and compare it both with experiments and with findings obtained for the thermally activated process.

If not otherwise stated, we use units such that $\hbar = 1 = k_B$ in this chapter. The bath is specified by the spectral density of the commutator (6.17,6.19). It proves to be convenient to define the spectrum of the symmetrized correlation function of the strain field,

$$\tilde{J}(\omega) = \frac{1}{4}\int_{-\infty}^{\infty} dt\, e^{i\omega t} \langle f(t)f + ff(t)\rangle_B. \tag{7.1}$$

As in (6.17), thermal average and time evolution are performed with respect to H_B. By direct calculation or by applying relations such as (A10, A11), one finds

$$\tilde{J}(\omega) = J(\omega)\coth(\beta\omega/2) = \pi(\alpha/\tilde{\omega}^2)\omega^3\coth(\beta\omega/2) \tag{7.2}$$

where $\alpha/\tilde{\omega}^2$ is a phenomenological coupling constant containing the deformation potential, the mass density and the sound velocity.

In the sequel, time evolution of symmetrized correlation functions

$$G(t-t') = \tfrac{1}{2}\langle \sigma_z(t)\sigma_z(t') + \sigma_z(t)\sigma_z(t')\rangle, \tag{7.3}$$

$$C(t-t') = \tfrac{1}{2}\langle \sigma_x(t)\sigma_x(t') + \sigma_x(t)\sigma_x(t')\rangle, \tag{7.4}$$

is calculated from the Heisenberg equation of motion $\dot{A} = i[H, A]$,

$$\dot{\sigma}_x = -\Delta\sigma_y - f\sigma_y, \tag{7.5}$$

$$\dot{\sigma}_y = \Delta\sigma_x + f\sigma_x - \Delta_0\sigma_z, \tag{7.6}$$

$$\dot{\sigma}_z = \Delta_0\sigma_y. \tag{7.7}$$

Here, $\sigma_i(t) = e^{i\mathcal{L}t}\sigma_i = e^{iHt}\sigma_i e^{-iHt}$ denotes a spin-operator in the Heisenberg picture, and $\mathcal{L}* = [H, *]$ is the quantum mechanical Liouville operator.

Angular brackets $\langle \ldots \rangle$ indicate the thermal average with respect to the equilibrium density matrix $\rho_{\text{eq}} = \mathrm{e}^{-\beta H}\mathrm{tr}(\mathrm{e}^{-\beta H})$. All experimental information is contained in the corresponding spectral functions

$$G''(\omega) = \frac{1}{2}\int_{-\infty}^{\infty} \mathrm{d}t\, \mathrm{e}^{\mathrm{i}\omega t} G(t), \tag{7.8}$$

and $C''(\omega)$, defined analogously.

In order to account properly for the low-temperature relaxation behavior, we will introduce fluctuation operators whose correlations yield the time-dependent parts of (7.3) and (7.4),

$$\delta G(t) = G(t) - \langle \sigma_z \rangle^2, \tag{7.9}$$
$$\delta C(t) = C(t) - \langle \sigma_x \rangle^2. \tag{7.10}$$

In both weak and strong coupling regimes we will use the Mori-Zwanzig projection operator formalism [119,120]. Although a formulation of the dynamics in one projection scheme is possible [94], in the present context it turns out advantageous to apply different schemes for the two asymptotic regimes.

7.1 Coherent Tunneling: Perturbation Theory

In the low-temperature (or weak-coupling) limit, a formulation of the dynamics in the eigenbasis of the spin part of the Liouvillian is most convenient. For this purpose we define new spin operators

$$\sigma_0 = \frac{\Delta_0}{E}\sigma_x + \frac{\Delta}{E}\sigma_z, \tag{7.11}$$

$$\sigma_\pm = \frac{1}{\sqrt{2}}\left(\frac{\Delta_0}{E}\sigma_z - \frac{\Delta}{E}\sigma_x \mp \mathrm{i}\sigma_y\right), \tag{7.12}$$

which transform the Hamiltonian into

$$H = \frac{1}{2}E\sigma_0 + \frac{1}{2}f\left(\frac{\Delta}{E}\sigma_0 + \frac{1}{\sqrt{2}}\frac{\Delta_0}{E}(\sigma_+ + \sigma_-)\right) + H_{\mathrm{B}}. \tag{7.13}$$

In this basis, H is diagonal in the limit of vanishing spin-phonon coupling ($g \to 0$). The operators (7.11, 7.12) satisfy the commutation relations $[\sigma_0, \sigma_\pm] = \pm 2\sigma_\pm$ and $[\sigma_+, \sigma_-] = \sigma_0$.

In order to deal properly with the static part of the correlation function $G(t)$, we project the dynamics onto the space spanned by the spin-fluctuation operators ($i = 0, \pm$)

$$\delta\sigma_i = \sigma_i - \langle \sigma_i \rangle. \tag{7.14}$$

These operators span an orthogonal basis

$$(\delta\sigma_i|\delta\sigma_j) \equiv \eta_{ij} = \eta_{ii}\delta_{ij} \tag{7.15}$$

with respect to the scalar product $(A|B) \equiv \frac{1}{2}\langle A^\dagger B + BA^\dagger \rangle$. When we take $\langle \sigma_\pm \rangle = 0$, we find for the static correlations

$$\eta_{00} = 1 - \langle \sigma_0 \rangle^2, \qquad \eta_{\pm\pm} = 1. \tag{7.16}$$

With

$$\delta C_{ij}(t) = (\delta\sigma_i(t)|\delta\sigma_j) \tag{7.17}$$

the complex correlation matrix $(i, j = 0, \pm)$

$$\delta C_{ij}(z) = i \int_0^\infty dt\, e^{izt} \delta C_{ij}(t) \qquad \text{Im } z > 0 \tag{7.18}$$

can be expressed as a resolvent matrix element

$$\delta C_{ij}(z) = (\delta\sigma_i | [\mathcal{L} - z]^{-1} | \delta\sigma_j). \tag{7.19}$$

Mori's reduction scheme can now be performed by defining the projector

$$\mathcal{P} = \sum_{i=0,\pm} |\delta\sigma_i) \eta_{ii}^{-1} (\delta\sigma_i| = \mathcal{I} - \mathcal{Q} \tag{7.20}$$

and applying the resolvent identity (5.13) onto (7.19)

Since $\mathcal{P}$, $\mathcal{Q}$, and $\mathcal{I}$ are three-dimensional projections, the continued fraction expansion involves matrices. When we truncate at lowest order, we obtain the exact matrix equation

$$\delta \mathbf{C}(z) = \eta \frac{-1}{z\eta - \Omega + \mathbf{M}(z)} \eta, \tag{7.21}$$

where the frequency matrix

$$i\Omega_{ij} = \delta\dot{C}_{ij}(t=0) \tag{7.22}$$

accounts for the free spin dynamics, and the memory matrix

$$M_{ij}(t) = (\mathcal{Q}\mathcal{L}\delta\sigma_i | e^{-i\mathcal{Q}\mathcal{L}\mathcal{Q}t} \mathcal{Q}\mathcal{L}\delta\sigma_j) \tag{7.23}$$

for the influence of the bath on the spin motion. The former yields, in a straightforward fashion,

$$\Omega = \begin{pmatrix} 0 & 0 & 0 \\ 0 & E & 0 \\ 0 & 0 & -E \end{pmatrix}. \tag{7.24}$$

The derivatives appearing in the memory matrix are easily calculated. Since the operators $\mathcal{Q}\mathcal{L}\delta\sigma_i$ are already linear in the coupling constants λ_k, the lowest-order Born approximation is achieved by replacing $\mathcal{Q}\mathcal{L}\mathcal{Q}$ with the free spin dynamics $\mathcal{L}_0 * = (1/2)E[\sigma_0, *]$ and calculating thermal expectation values with respect to $H_0 = (1/2)E\sigma_0$. Thereby, one finds that the spin and the bath dynamics factorize and that $\langle \sigma_0 \rangle = -\tanh(\beta E/2)$ and $\eta_{00} = 1/\cosh^2(\beta E/2)$. With the definition

$$m_i(t) = (\sigma_i(t) f(t) | \sigma_i f)_0, \tag{7.25}$$

the memory function reads in first Born approximation

$$\mathbf{M} = \begin{pmatrix} 2r(m_+ + m_-) & 0 & 0 \\ 0 & \frac{1}{2}rm_0 + (1-r)m_+ & -\frac{1}{2}rm_0 \\ 0 & -\frac{1}{2}rm_0 & \frac{1}{2}rm_0 + (1-r)m_- \end{pmatrix} \quad (7.26)$$

where $r \equiv \Delta_0^2/E^2$. The Fourier transform of $m_i(t)$ is given by the weighted convolution integral of free spin spectral functions

$$C_{00}^{0\prime\prime}(\omega) = \pi\delta(\omega), \quad (7.27)$$

$$C_{\pm\pm}^{0\prime\prime}(\omega) = \pi\delta(\omega \mp E), \quad (7.28)$$

with the spectral function of the bath (7.2),

$$m_i''(\omega) = \left(C_{ii}^{0\prime\prime}(\omega) * \widetilde{J}(\omega)\right), \quad (7.29)$$

where we have used the weighted convolution integral (A13) defined in Appendix A. This convolution is easily performed; after evaluating the frequency-dependent memory functions $m_i''(\omega)$ at the bare poles $\omega = 0, \pm E$, one finds the damping constants

$$\gamma = \eta_{00}^{-1}\left(rm_+''(0) + rm_-''(0)\right) \quad (7.30)$$

$$\Gamma = \frac{1}{2}\eta_{\pm\pm}^{-1}\left(rm_0''(E) + (1-r)m_\pm''(\pm E)\right), \quad (7.31)$$

which read with (7.29) and (6.13)

$$\gamma \equiv 2\Gamma = \pi\frac{\alpha}{\widetilde{\omega}^2}\Delta_0^2 E \coth(\beta E/2). \quad (7.32)$$

By using again the transformation (7.11, 7.12), we calculate the residues of the poles $z_\pm = \pm\widetilde{E} - i\Gamma$, $z_0 = -i\gamma$ up to $O(\gamma/E)$ and $O(\Gamma/E)$, where $\widetilde{E} = \sqrt{E^2 - \Gamma^2}$. The resulting spectral function

$$G''(\omega) = \pi\langle\sigma_z\rangle^2\delta(\omega) + \delta G''(\omega) \quad (7.33)$$

comprises in particular a delta peak at zero frequency; the remainder

$$\delta G''(\omega) = \frac{\Delta^2/E^2}{\cosh^2(\beta E/2)}\frac{\gamma}{\omega^2 + \gamma^2} + \frac{1}{2}\frac{\Delta_0^2}{E^2}\sum_\pm \frac{\Gamma}{(\omega \pm \widetilde{E})^2 + \Gamma^2} \quad (7.34)$$

contains one quasi-elastic and two inelastic resonances with finite widths.

When performing the Fourier back-transformation, the undamped part of the spectrum gives rise to a finite value for the correlation function in the long-time limit,

$$G(t \to \infty) = \langle\sigma_z\rangle^2 = \frac{\Delta^2}{E^2}\tanh^2(\beta E/2). \quad (7.35)$$

Absorption spectroscopy probes the fluctuation spectrum $\delta G''(\omega)$ instead. The finite widths lead to an exponential damping in time representation,

$$\delta G(t) = \frac{\Delta^2/E^2}{\cosh^2(\beta E/2)} e^{-\gamma t} + \frac{\Delta_0^2}{E^2} e^{-\Gamma t} \cos(\widetilde{E}t - \phi)/\cos(\phi), \qquad (7.36)$$

where $\tan(\phi) = \Gamma/\widetilde{E}$.

Equation (7.36) is equivalent to Jäckle's result obtained from a Boltzmann equation approach [105].

7.2 Incoherent Tunneling: Mode-Coupling Theory

When we calculate higher-order terms of the perturbation series for the widths γ and Γ, we find that the perturbative approach breaks down at a temperature where the quantity $(\alpha/\widetilde{\omega}^2)T^2$ is of the order of unity. The rates then increase strongly with temperature; the spectral lines of $G''(\omega)$ broaden and move towards the central peak. For the symmetric case $\Delta = 0$, it has been shown that in the scheme of a mode-coupling approximation the two inelastic resonances of $G''(\omega)$ merge into one single quasi-elastic resonance at $(\alpha/\widetilde{\omega}^2)T^2 \approx 1$ whose width narrows with further increasing temperatures [95]. This picture remains essentially unchanged in the biased case $\Delta \neq 0$.

In the incoherent regime, the dynamics is most easily formulated in terms of the spin-operators σ_x, σ_y and σ_z. Because correlations of different σ_i become less important with increasing temperatures, a continued fraction representation of the resolvents

$$G(z) = (\sigma_z|[\mathcal{L} - z]^{-1}|\sigma_z), \qquad (7.37)$$
$$C(z) = (\sigma_x|[\mathcal{L} - z]^{-1}|\sigma_x), \qquad (7.38)$$

is appropriate for determining the dynamics of (7.3) and (7.4). In the sequel we will repeat the projection procedure for both resolvents up to the stage at which all memory functions are built by spin-bath operators $f\sigma_i$.

Longitudinal Correlation Function $C(z)$

When we apply Mori's reduction procedure with the projector $\mathcal{P}_x = |\sigma_x)(\sigma_x|$ on (7.38), we obtain

$$C(z) = \frac{-1}{z + N(z)}, \qquad (7.39)$$

where the memory function

$$N(z) = (\mathcal{Q}_x\dot{\sigma}_x|[\mathcal{Q}_x\mathcal{L} - z]^{-1}|\mathcal{Q}_x\dot{\sigma}_x) \qquad (7.40)$$

is defined by the operator

$$\mathcal{Q}_x\dot{\sigma}_x = -\Delta\sigma_y - f\sigma_y. \qquad (7.41)$$

Now we separate $N(z)$ into two parts

$$N(z) = \Delta^2 \hat{Y}(z) + N_1(z), \tag{7.42}$$

the first one arising from $\Delta \sigma_y$ in (7.41),

$$\hat{Y}(z) = (\sigma_y |[\mathcal{Q}_x \mathcal{L} - z]^{-1}| \sigma_y), \tag{7.43}$$

and $N_1(z)$ containing the remainder. The latter embraces three terms, each of which involves the composite operator $f\sigma_y$. Thus we keep $N_1(z)$ as it stands, and repeat the projection for $\hat{Y}(z)$,

$$\hat{Y}(z) = \frac{-1}{z + \hat{N}(z)}, \tag{7.44}$$

with the corresponding memory function

$$\hat{N}(z) = (\mathcal{Q}_y \mathcal{Q}_x \dot{\sigma}_y |[\mathcal{Q}_y \mathcal{Q}_x \mathcal{L} - z]^{-1}| \mathcal{Q}_y \mathcal{Q}_x \dot{\sigma}_y) \tag{7.45}$$

involving the operator

$$\mathcal{Q}_y \mathcal{Q}_x \dot{\sigma}_y = f\sigma_x - \Delta_0 \sigma_z. \tag{7.46}$$

When we split off the part arising from $\Delta_0 \sigma_z$,

$$\hat{G}(z) = (\sigma_z |[\mathcal{Q}_y \mathcal{Q}_x \mathcal{L} - z]^{-1}| \sigma_z), \tag{7.47}$$

we obtain $\hat{N}(z)$ as a sum of two terms

$$\hat{N}(z) = \Delta_0^2 \hat{G}(z) + M_1(z), \tag{7.48}$$

where as above $M_1(z)$ contains the remainder. Due to $\mathcal{Q}_z \mathcal{Q}_y \mathcal{Q}_x \dot{\sigma}_z = 0$ the correlation function $\hat{G}(z)$ has no dynamics and is therefore given by $\hat{G}(z) = -1/z$.

Thus we have derived a continued fraction representation for the longitudinal correlation function (7.38),

$$C(z) = \frac{-1}{z + N_1(z) - \dfrac{\Delta^2}{z + M_1(z) - \dfrac{\Delta_0^2}{z}}}. \tag{7.49}$$

Transverse Correlation Function $G(z)$

The different equation of motion causes a few essential modifications as compared to $C(z)$; hence we shortly repeat the above scheme. When we project first on σ_z and then on σ_y, we get for (7.37)

$$G(z) = \frac{-1}{z - \dfrac{\Delta_0^2}{z + M(z)}}; \tag{7.50}$$

with the memory function

$$M(z) = (\mathcal{Q}_y \mathcal{Q}_z \mathcal{L} \sigma_y |[\mathcal{Q}_y \mathcal{Q}_z \mathcal{L} - z]^{-1}| \mathcal{Q}_y \mathcal{Q}_z \mathcal{L} \sigma_y) \tag{7.51}$$

defined by the operator
$$i\mathcal{Q}_y\mathcal{Q}_z\mathcal{L}\sigma_y = \Delta\sigma_x + \sigma_x e. \tag{7.52}$$
When we split off the part of $M(z)$ which arises from the first term in (7.52),
$$M(z) = \Delta^2 \hat{C}(z) + M_2(z) \tag{7.53}$$
with
$$\hat{C}(z) = (\sigma_x|[\mathcal{Q}_y\mathcal{Q}_z\mathcal{L} - z]^{-1}|\sigma_x), \tag{7.54}$$
and apply the resolvent identity once more on $\hat{C}(z)$,
$$\hat{C}(z) = \frac{-1}{z + N_2(z)}, \tag{7.55}$$
we obtain with $\mathcal{Q} \equiv \mathcal{Q}_x\mathcal{Q}_y\mathcal{Q}_z$ and
$$\mathcal{Q}\dot{\sigma}_x = -e\sigma_y \tag{7.56}$$
the second memory function
$$N_2(z) = (\mathcal{Q}\dot{\sigma}_x|[\mathcal{Q}\mathcal{L} - z]^{-1}|\mathcal{Q}\dot{\sigma}_x). \tag{7.57}$$
Both $M_2(z)$ and $N_2(z)$ involve composite operators $f\sigma_i$. At this point we break off the continued fraction; as a result we obtain a representation for the transverse correlation function similar to (7.38),
$$G(z) = \frac{-1}{z - \dfrac{\Delta_0^2}{z + M_2(z) - \dfrac{\Delta^2}{z + N_2(z)}}}. \tag{7.58}$$

These continued fractions clearly show the difficulties arising from a finite bias $\Delta \neq 0$. Whereas in the symmetric case $C(z)$ has one pole and $G(z)$ has two poles, here both functions have a three-pole structure. This becomes evident by writing (7.49, 7.58) as
$$C(z) = -\frac{z(z + M_1(z)) - \Delta_0^2}{(z + N_1(z))(z^2 + zM_1(z) - \Delta_0^2) - z\Delta^2}, \tag{7.59}$$
$$G(z) = -\frac{(z + M_2(z))(z + N_2(z)) - \Delta^2}{(z + N_2(z))(z^2 + izM_2(z) - \Delta_0^2) - z\Delta^2}. \tag{7.60}$$
The memory functions $M_\alpha(z)$ and $N_\alpha(z)$ ($\alpha = 1, 2$) contain auto-correlations between spin-bath operators $f\sigma_i$, and correlations between spin-bath $f\sigma_i$ and spin operators σ_j, furnished with complicated projected $\mathcal{Q}\mathcal{L}$-dynamics.

Mode-Coupling Approximation

The projected dynamics in the memory functions cannot be treated exactly. In a mode-coupling approximation, correlation functions of products of operators are decoupled into products of single-operator correlation functions, and the projected $Q\mathcal{L}$-dynamics is replaced by the full $\mathcal{L}$-dynamics. Thereby the dynamic equations are closed and the correlation functions can be calculated self-consistently.

In the present case the decoupling approximation for composite operators amounts to

$$M_1(t) = M_2(t) = \frac{1}{2}\sum_{\pm}\langle\sigma_x(\pm t)\sigma_x\rangle\langle f(\pm t)f\rangle \equiv M(t), \tag{7.61}$$

$$N_1(t) = N_2(t) = \frac{1}{2}\sum_{\pm}\langle\sigma_y(\pm t)\sigma_y\rangle\langle f(\pm t)f\rangle \equiv N(t). \tag{7.62}$$

Because of the zero expectation value of the bath operator, $\langle f\rangle = 0$, mixed correlation functions vanish when we take $\langle\sigma_i(t)\sigma_j f\rangle \to \langle\sigma_i(t)\sigma_j\rangle\langle f\rangle$.

By using $\partial_t^2\langle\sigma_z(t)\sigma_z\rangle = -\Delta_0^2\langle\sigma_y(t)\sigma_y\rangle$ and taking the Fourier transform, we find the spectral functions

$$M''(\omega) = (C''(\omega) * \widetilde{J}(\omega)), \tag{7.63}$$
$$N''(\omega) = ((\omega/\Delta_0)^2 G''(\omega) * \widetilde{J}(\omega)), \tag{7.64}$$

where the convolution product is defined in (A13). In principle, the reactive parts $M'(\omega)$ and $N'(\omega)$ can be determined from the spectra $M''(\omega)$ and $N''(\omega)$ via a Kramers-Kronig relation. Together with (7.49, 7.58), these equations form a closed set of non-linear, self-consistent equations for $G(z)$ and $C(z)$ or $M(z)$ and $N(z)$. A numerical solution has shown a transition from coherent to incoherent tunneling occurring analogously to the symmetric case [94,95]. To make this quantitative we present an approximative analytical solution for the mode-coupling equations (7.49, 7.58, 7.63, 7.64).

Pole Approximation

A numerical solution of the mode-coupling equations shows that, in the strongly overdamped regime, the spectrum of $G(t)$ consists of one single peak centered at $\omega = 0$. The width narrows with increasing temperatures and the line shape approaches a Lorentzian form, i.e., the memory functions (7.63, 7.64) become less frequency dependent. This justifies a pole approximation at $\omega = 0$ in which $M''(\omega)$ and $N''(\omega)$ are set equal,

$$M''(0) = N''(0) \equiv \widetilde{\Gamma}. \tag{7.65}$$

In this approximation the reactive parts $M'(\omega)$ and $N'(\omega)$ vanish since they are odd in ω. (Concerning this point, see the discussion at the beginning of Chap. 8, p. 153.)

7.2 Incoherent Tunneling: Mode-Coupling Theory

The poles of $G(z)$ and $C(z)$ are determined by the complex roots of the cubic equation

$$(z + i\widetilde{\Gamma})(z^2 + iz\widetilde{\Gamma} - \Delta_0^2) - z\Delta^2$$
$$\equiv (z + i\gamma)((z + i\Gamma)^2 - \widetilde{E}^2) + R(\widetilde{\Gamma}) = 0, \quad (7.66)$$

which defines the quantities

$$\gamma = \widetilde{\Gamma}\frac{\Delta_0^2}{E^2 + \widetilde{\Gamma}^2}, \qquad \Gamma = \widetilde{\Gamma} - \tfrac{1}{2}\gamma, \quad (7.67)$$

$$\widetilde{E} = \sqrt{E^2 - \Gamma^2 + (\widetilde{\Gamma} - \gamma)^2}, \quad (7.68)$$

$$R(\widetilde{\Gamma}) = \Delta_0^2\widetilde{\Gamma} - \gamma\left(E^2 + \widetilde{\Gamma}^2 + \gamma(\gamma - 2\widetilde{\Gamma})\right). \quad (7.69)$$

It is easily seen that $R(\widetilde{\Gamma})$ is negligible in both asymptotic regimes $\widetilde{\Gamma} \ll E$ and $\widetilde{\Gamma} \gg E$. The cubic equation (7.66) is essentially identical to the pole equation arising from a Ohmic heat bath at high temperatures which has been derived in [16]. In the present case, however, the width $\widetilde{\Gamma}$ shows quite a different temperature dependence. In the sequel we neglect the term $R(\widetilde{\Gamma})$.

Thus the roots of $G(z)$ and $C(z)$ are approximately given by $z_0 = -i\gamma$ and $z_\pm = \pm\widetilde{E} - i\Gamma$; in the strongly overdamped regime, we have

$$\gamma = \Delta_0^2/\widetilde{\Gamma}, \qquad \Gamma = \widetilde{\Gamma}, \qquad \widetilde{E} = \Delta \qquad \text{for} \quad \widetilde{\Gamma} \gg E. \quad (7.70)$$

In this limit the residue of the purely imaginary pole z_0 of $G(z)$ approaches unity whereas the residues of the oscillating poles $z_\pm$ vanish, and vice versa for $C(z)$. Thus we find for $\widetilde{\Gamma} \gg E$ the spectra

$$G''(\omega) \approx \frac{\Delta_0^2/\widetilde{\Gamma}}{\omega^2 + \Delta_0^4/\widetilde{\Gamma}^2}, \quad (7.71)$$

$$C''(\omega) \approx \frac{\widetilde{\Gamma}}{\omega^2 + \widetilde{\Gamma}^2}. \quad (7.72)$$

By performing Fourier back-transformation, we obtain the correlation functions

$$G(t) \approx e^{-\Delta_0^2 t/\widetilde{\Gamma}}, \quad (7.73)$$

$$C(t) \approx e^{-\widetilde{\Gamma}t}. \quad (7.74)$$

The temperature dependence of the rate $\widetilde{\Gamma}$ remains to be determined. For this purpose we insert the expression (7.71) for $G''(\omega)$ into (7.64) and evaluate the convolution at $\omega = 0$. This yields a self-consistent equation for the rate $\widetilde{\Gamma}$,

$$\widetilde{\Gamma} = \frac{4}{\widetilde{\Gamma}}\frac{\alpha}{\widetilde{\omega}^2}\int_0^\infty \frac{\Omega^5}{\Omega^2 + \Delta_0^4/\widetilde{\Gamma}^2}\frac{d\Omega}{\sinh(\beta\Omega)}. \quad (7.75)$$

Since $T \gg \Delta_0^2/\widetilde{\Gamma}$, we neglect $\Delta_0^4/\widetilde{\Gamma}^2$ in the denominator, which permits the integration of (7.75) and gives

$$\widetilde{\Gamma} = \frac{\pi^2}{\sqrt{2}} \sqrt{\frac{\alpha}{\widetilde{\omega}^2}} T^2. \tag{7.76}$$

By inserting (7.72) into (7.63), we obtain the same expression for $\widetilde{\Gamma}$, thus affirming the consistency of our initial assumption (7.65) and completing the solution of the mode-coupling equations in the incoherent regime. In [95,97] the self-consistency relations have been solved in more detail, with results identical to those given here.

Relaxation of Asymmetric Tunneling Systems

Here we reinsert $\hbar$ and k_B. In the two proceeding sections we have solved the dynamics of an asymmetric TS in the two asymptotic regimes of underdamped and overdamped motion by applying a perturbative and a mode-coupling scheme, respectively. In the intermediate regime where $\hbar\widetilde{\Gamma} \approx E$, neither approach is expected to be valid and no analytical result is available for the biased case. (Being based on a more elaborate ansatz for the memory functions, the solution for the symmetric case $\Delta = 0$ given in [95] covers the whole parameter range.)

However, for all practical purposes it is sufficient to match the results for the two asymptotic regimes appropriately. Since at high temperatures $k_B T \gg E$ the static spin polarization $\langle \sigma_i \rangle$ is negligible, we can identify $\delta G(z)$ and $G(z)$ in the incoherent regime. Then the following formulae reasonably interpolate between the behavior in the coherent (7.32, 7.34) and the incoherent regime (7.71, 7.76):

$$\delta G''(\omega) = \frac{\hbar^2 \widetilde{\Gamma}^2 + \Delta^2 \cosh(\beta E/2)^{-2}}{\hbar^2 \widetilde{\Gamma}^2 + E^2} \frac{\tau_1}{1 + \omega^2 \tau_1^2}$$
$$+ \frac{1}{2} \frac{\Delta_0^2}{\hbar^2 \Gamma^2 + E^2} \sum_{\pm} \frac{\tau_2}{1 + \tau_2^2 (\omega \pm \widetilde{E}/\hbar)^2}. \tag{7.77}$$

The relaxation time $\tau_1 = 1/\gamma$ is given by

$$\gamma = \widetilde{\Gamma} \frac{\Delta_0^2}{\hbar^2 \widetilde{\Gamma}^2 + E^2} \equiv \frac{\hbar r}{\tau_{\min}}, \tag{7.78}$$

and the phase coherence time $\tau_2 = /\Gamma$ by

$$\Gamma = \begin{cases} r\widetilde{\Gamma} & T < T^* \\ \widetilde{\Gamma} & T \geq T^* \end{cases} \tag{7.79}$$

($\tau_{\min}$ is independent of $r \equiv \Delta_0^2/E^2$.) The quantity $\widetilde{\Gamma}$ reads in the two regimes

$$\widetilde{\varGamma} = \begin{cases} (1/2\pi\hbar)\widetilde{\gamma}^2 E^3 \coth(\beta E/2) \equiv \widetilde{\varGamma}_{\mathrm{1ph}} & T < T^* \\ (\pi/2\hbar)\widetilde{\gamma}\,(k_{\mathrm{B}}T)^2 \equiv \widetilde{\varGamma}_{\mathrm{MC}} & T \geq T^*. \end{cases} \quad (7.80)$$

Here we have expressed the phenomenological coupling constant $\alpha/\widetilde{\omega}^2$ through the mass density ϱ, the sound velocity v and the deformation potential [124] g, according to

$$2\alpha/\widetilde{\omega}^2 \equiv \hbar^2 \widetilde{\gamma}^2 = g^2/(\varrho v^5 \hbar). \quad (7.81)$$

At the temperature T^* the dynamics changes drastically. T^* is defined by $\hbar\widetilde{\varGamma} = E$ or, for thermal two-level systems with $E \approx k_{\mathrm{B}}T$, by $\hbar\widetilde{\varGamma}_{\mathrm{MC}}(T^*) \equiv k_{\mathrm{B}}T^*$, yielding

$$k_{\mathrm{B}}T^* = \frac{2}{\pi}\frac{1}{\widetilde{\gamma}}. \quad (7.82)$$

In glasses one finds TLSs with a wide distribution of level splittings E; at temperature T^* all thermal TLSs ($E \leq k_B T$) are overdamped. This is equivalent to the condition $(k_{\mathrm{B}}^2/\hbar^2)(\alpha/\widetilde{\omega}^2)T^{*2} \approx 1$ derived for the transition to incoherent motion in the symmetric case (see Ref. [95]).

7.3 Dynamics Beyond the Two-Level Approximation

Damping of tunneling systems in dielectric glasses has been mainly treated using perturbation theory with respect to the elastic coupling [105,110,115]. In the two-state approximation (6.16), phonon scattering occurs between the two ground-state levels separated by the energy E; perturbation theory yields the one-phonon rate [105]

$$\frac{1}{\tau_{\mathrm{1Ph}}} = \frac{1}{2\pi}\frac{g^2}{\varrho v^5 \hbar^4}\Delta_0^2 E \coth\left(\frac{E}{2k_{\mathrm{B}}T}\right); \quad (7.83)$$

g and v are properly averaged over the longitudinal and transverse phonon branches. Above 5 K, sound velocity and sound attenuation exhibit a significant change with respect to temperature dependence (see Figs. 6.2–6.4), which are not accounted for by the one-phonon rate (7.83). In order to explain these deviations, two different scenarios have been proposed, namely (i) incoherent tunneling arising from overdamped two-level systems and (ii) the break-down of the two-level description. It is not obvious from the beginning which of these scenarios provides the proper description for tunneling dynamics above 5 K; in the sequel of this chapter we discuss both approaches in close comparison with experimental data for various oxide glasses.

(a) Mode-coupling theory indicates that finite-order perturbation theory breaks down above a temperature T^* [see (7.82)], resulting in incoherent tunneling motion with relaxation rate

$$\frac{1}{\tau_{\mathrm{MC}}} = \frac{\pi}{2}\sqrt{\frac{g^2 k_{\mathrm{B}}^4}{\varrho v^5 \hbar^5}}T^2. \quad (7.84)$$

This picture still relies on the two-state description; yet the thermal motion strongly influences the tunneling dynamics and destroys the phase coherence of quantum oscillations.

(b) With rising temperature, excited states in the double-well potential become important and contribute a term of the form const.$\times e^{-E_1/k_B T}$ to the damping rate (sometimes referred to as the Orbach process), with the excitation energy E_1; under certain circumstances, taking the sum over all higher levels $E_1, E_2, E_3, \ldots$ results for $\Delta \ll k_B T$ in the rate

$$\frac{1}{\tau_{\text{th}}} = \frac{1}{\tau_0} \exp\left(-\frac{V}{k_B T}\right), \tag{7.85}$$

where $1/\tau_0$ is of the order of the Debye frequency and V is the barrier height between the two wells. In any case, because of its activated behavior, such a rate will at some point exceed the quantum tunneling rate (7.83) which varies only linearly with T [143,147,145,144,146]. (Due to the large prefactor $1/\tau_0$, this cross-over will occur at a temperature well below the barrier height V.)

Since the seminal work of Kramers [160], there has been much progress made in the understanding of the transition between the tunneling and the thermally activated regime (see [20]). These works are mainly concerned with a modification of the pre-exponential factor, the so-called attempt frequency τ_0^{-1}. However, with respect to the ensemble average to be performed later, the attempt frequency only contributes logarithmically and the modification mentioned plays no essential role in glasses. Thus, we may safely treat τ_0^{-1} as a constant that is of the order of the Debye frequency.

Finally we mention an extension of the tunneling model put forward by Klinger and his co-workers [147] and popularized recently by Parshin, Buchenau, and others [145,144,146]. This so-called soft-potential model (SPM) provides a description for the low-temperature properties of glasses in terms of a quartic potential with an appropriate distribution for its three parameters (see [146]). Whereas at very low temperatures it reduces essentially to the tunneling model, with rising temperatures local quasi-harmonic modes (arising from quartic potentials with a vanishing or positive quadratic term) become important; with these localized modes, the SPM qualitatively accounts for universal properties of glasses at higher temperatures (above ca. 1 K) – like the plateau in the thermal conductivity [144] and the bump in the specific heat [145]. In the present context, the SPM is equivalent to the tunneling model including thermal activation.

7.4 Sound Propagation

In glasses one finds an ensemble of tunneling systems (TSs) with a wide distribution of parameters, as given in (6.11–6.14). In order to compare our calculation with experiments, an ensemble average over all TSs has to be performed. In the following we will always assume that the upper bound $E_{\max}$ is much larger than all other energies; hence $E_{\max}$ does not appear explicitly in our results. Macroscopic quantities are obtained by the average

$$\overline{\mathcal{O}} = \int_0^{E_{\max}} dE \int_{r_{\min}}^1 dr\, P(E,r)\mathcal{O}. \tag{7.86}$$

Acoustic or microwave spectroscopy probes the time evolution of the operator σ_z. For weak external fields, linear-response theory is applicable; therefore any experimental quantity can be expressed in terms of the symmetrized transverse correlation spectrum $\delta G''(\omega)$. In particular, it is related to the dynamic susceptibility

$$\chi(z) = \frac{i}{\hbar}\int_0^\infty dt\, e^{izt}\langle[\sigma_z(t),\sigma_z]\rangle \tag{7.87}$$

by a fluctuation-dissipation theorem (FDT)

$$\chi''(\omega) = \frac{2}{\hbar}\tanh(\beta\hbar\omega/2)\delta G''(\omega), \tag{7.88}$$

where $\chi(z = \omega + i0^+) = \chi'(\omega) + i\chi''(\omega)$. Experimentally accessible quantities, like internal friction and variation of sound velocity, follow from this via [107]

$$Q^{-1} = \frac{g^2}{\varrho v^2}\overline{\chi''(\omega)}, \tag{7.89}$$

$$\frac{\delta v}{v} = -\frac{1}{2}\frac{g^2}{\varrho v^2}\overline{\chi'(\omega)}. \tag{7.90}$$

Here a word concerning notation is in order. We follow the general usage when denoting the commutator correlation function of a single tunneling system by χ. (The dimension of the Laplace transform given in (7.87) is inverse energy.) The relations with sound velocity and attenuation, (7.89, 7.90) assume $\chi(z)$ to be identical to the elastic response function.

This is incompatible with the notation used in Chaps. 2–5, where the dielectric susceptibility of a single impurity has been denoted by $\alpha(\omega)$, and the elastic response function by $R(\omega)$. The letter χ has been reserved for the averaged, or macroscopic, dielectric susceptibility (which is dimensionless.)

In Chap. 2 we found the elastic and dielectric response functions of a lithium impurity to be significantly different. Yet tunneling systems in glasses have been found to respond in a similar fashion to acoustic and electromagnetic fields. In any case, the present two-state approximation would not allow for a proper distinction.

The relaxation process of biased TSs at sufficiently low temperatures has been elaborated by Jäckle [105] using Fermi's Golden Rule. The resulting relaxation rates are given by the one-phonon expression (7.32) and the dynamics is governed by coherent tunneling, cf. (7.34).

The relaxation of overdamped TSs has not been considered so far. It was commonly assumed that thermally activated relaxation was responsible for the experimentally observed deviation above 5 K from the prediction of the tunneling model. This was worked out in detail by Tielbürger et al. [143], and within the SPM by Buchenau et al. [144]. Before we summarize this theory in Sect. 7.4, let us show how far we can proceed within the two-level approximation by including the incoherent tunneling regime.

Damping by Two-Level Systems

We assume the two-level approximation to be valid beyond 5 K and we evaluate sound velocity and attenuation of elastic waves as they arise from scattering by two-level tunneling systems. For this purpose we have to perform the ensemble average over all TSs present in a sample. In linear response theory relaxational and resonant contributions may be calculated separately and then added.

The so-called resonant part of the elastic response function arises from the second term of (7.77); its evaluation will be restricted to the low-temperature or weak-coupling limit, since in the opposite case this contribution disappears. When we insert (7.77) in (7.88), assume $\hbar\widetilde{\Gamma} \ll E$, and take the average over E and r, we obtain with $\int_0^1 dr(1-r)^{-1/2} = 2$ the resonant absorption

$$Q^{-1}|_{\text{res}} = \pi C \tanh(\beta\hbar\omega/2), \tag{7.91}$$

where we have defined the dimensionless constant

$$C = \frac{\overline{P}g^2}{\varrho v^2}. \tag{7.92}$$

Except for very low temperatures and high frequencies, (7.91) is negligible because of the temperature factor, which in general is small for acoustic frequencies, $\tanh(\beta\hbar\omega/2) \ll 1$.

For the resonant contribution to the change of sound velocity, one obtains in the limit of small frequency $\beta\hbar\omega \ll 1$ [105,110]

$$\delta v/v|_{\text{res}} = C \ln(T/T_0), \tag{7.93}$$

where T_0 is an arbitrary reference temperature. According to mode-coupling theory, the condition $\hbar\widetilde{\Gamma} \ll E$ does not hold true for temperatures above 5 K. Thus the validity of the results for resonant interaction, (7.91, 7.93), is restricted to $T < T^*$.

Now we consider how relaxational motion of TSs affects a sound wave traveling through the glass. Accordingly, only the first part of (7.77) with the relaxational pole at $\omega = -i\gamma$ is relevant. For low-frequency acoustic experiments on glasses, the temperature exceeds by far the applied frequency, $\hbar\omega \ll k_B T$. Thus we may replace $\tanh(\beta\hbar\omega/2) \approx \beta\hbar\omega/2$ in the FDT which yields together with the Kramers-Kronig relation

$$\chi''(\omega) = \beta\omega\,\delta G''(\omega), \tag{7.94}$$

$$\chi'(\omega) = \beta\tau_1^{-1}\,\delta G''(\omega). \tag{7.95}$$

Using (6.14, 7.86, 7.89, 7.90) and the relations for the rates of Sect. 7.2, we find

$$Q^{-1} = C\int dE \int_{r_{\min}}^{1} dr\, F(E,r)\,\frac{1}{2k_\mathrm{B}T}\,\frac{\omega\tau_{\min}}{r^2 + \omega^2\tau_{\min}^2}, \tag{7.96}$$

$$\frac{\delta v}{v} = -C\int dE \int_{r_{\min}}^{1} dr\, F(E,r)\,\frac{1}{4k_\mathrm{B}T}\,\frac{r}{r^2 + \omega^2\tau_{\min}^2}, \tag{7.97}$$

where we have defined the function

$$F(E,r) = \frac{\widetilde{\Gamma}^2}{\widetilde{\Gamma}^2 + E^2}\,\frac{1}{\sqrt{1-r}} + \frac{E^2}{\widetilde{\Gamma}^2 + E^2}\,\frac{\sqrt{1-r}}{\cosh(\beta E/2)}. \tag{7.98}$$

Relaxation is most efficient for asymmetric tunneling systems, with $r \ll 1$; thus we may drop the factor $\sqrt{1-r}$ in the integrals. Then an analytical evaluation is possible for both limits $\omega\tau_{\min} \gg 1$ and $\omega\tau_{\min} \ll 1$. The temperature $\widetilde{T}$ is defined through the condition $\omega\tau_{\min} = 1$. Provided that the one-phonon process dominates at low temperatures this yields with (7.78, 7.80) and after substituting $E = k_\mathrm{B}T$,

$$k_\mathrm{B}\widetilde{T} = \left(\frac{\hbar\omega\pi}{\widetilde{\gamma}^2}\right)^{1/3}. \tag{7.99}$$

Cutting the E-integration at $E = \max(k_\mathrm{B}T, \hbar\widetilde{\Gamma})$, we find for the internal friction

$$Q^{-1} = \begin{cases} \pi^3 C\widetilde{\gamma}^2 k_\mathrm{B}^3 T^3/(24\hbar\omega) & \text{for } T < \widetilde{T} \\[6pt] \frac{\pi}{2}C\Big[1 - (\hbar\widetilde{\Gamma}_\mathrm{MC}/2k_\mathrm{B}T)\arctan\left(2k_\mathrm{B}T/\hbar\widetilde{\Gamma}_\mathrm{MC}\right) \\ \quad + (\hbar\widetilde{\Gamma}_\mathrm{MC}/2k_\mathrm{B}T)\left(\frac{\pi}{2} - \arctan\left(r_{\min}\widetilde{\Gamma}_\mathrm{MC}/\omega\right)\right)\Big] \\ \hfill \text{for } T > \widetilde{T}. \end{cases} \tag{7.100}$$

$\widetilde{\Gamma}_\mathrm{MC}$ is defined in (7.80). At very low temperatures $T < \widetilde{T}$ (corresponding to $\omega\tau_{\min} \gg 1$), relaxation attenuation increases with T^3; in the range $T > \widetilde{T}$ we distinguish three different laws,

$$Q^{-1} = \begin{cases} \frac{1}{2}\pi C & \widetilde{T} < T < T^* \\[4pt] \frac{1}{16}\pi^3 C\widetilde{\gamma}k_\mathrm{B}T & T^* < T < T_{\max} \\[4pt] \frac{1}{4}\pi C(\hbar\omega/r_{\min}k_\mathrm{B}T) & T > T_{\max}. \end{cases} \tag{7.101}$$

The absorption maximum occurs at temperature $T_{\max}$, which is obtained from (7.100) and the relation $\hbar\omega = r_{\min}\widetilde{\Gamma}_{\text{MC}}$,

$$k_B T_{\max} = \sqrt{\frac{2\hbar\omega}{\pi\widetilde{\gamma}r_{\min}}};\qquad(7.102)$$

the maximum temperature varies as $T_{\max} \propto \sqrt{\omega}$. Fig. 2. shows schematically the three characteristic temperatures $\widetilde{T}$, T^* and $T_{\max}$, and the typical lineshape of the internal friction.

The reactive part (7.97) can be integrated after discarding the factor $\sqrt{1-r}$; adding (7.93), one finds

$$\frac{\delta v}{v} = \begin{cases} C\log(T/T_0) & T < \widetilde{T} \\ -\frac{1}{2}C\log(T/T_0) & \widetilde{T} < T < T^* \\ -\left[\frac{1}{4} + \frac{\pi}{16}\right]C[T/T^*]\log(k_B T^2/\hbar\omega T^*) & T > T^*. \end{cases}\qquad(7.103)$$

The prefactor $-\frac{1}{2}$ of the logarithmic law for $\widetilde{T} < T < T^*$ is the sum of two terms, $+1$ from the resonant contribution (7.93) and $-\frac{3}{2}$ from the relaxational one.

Both the increase with T^3 and the plateau value $\pi C/2$ of the internal friction [105] and the logarithmic temperature dependence of the sound velocity [110,115] were derived more than 20 years ago. Below $\widetilde{T}$, even the fastest TSs (i.e. the symmetric TSs) are too slow to contribute to relaxation, so that for $T < \widetilde{T}$ the resonant interaction prevails. Below T^* one finds the well-known logarithmic temperature dependence of the sound velocity and the constant internal friction.

The novel features of the present work concern temperatures above T^*. At $T = T^*$, the temperature dependence changes to a linear increase in the absorption and a linear decrease in the sound velocity.

Damping by Activated Relaxation

Here we summarize the work of Tielbürger et al. [143] on thermally activated relaxation in glasses (see also [144] for a description within the soft-potential model). There it is assumed that the two-level description breaks down above 5 K as a consequence of a cross-over from the one-phonon rate (7.83) to a thermally activated one (7.85). According to these works, the tunneling dynamics changes from coherent to incoherent motion, and the increase in internal friction and the linear decrease in the sound velocity are explained by means of an appropriate distribution for the activation energy V.

As functions of tunneling amplitude Δ_0 and asymmetry energy Δ, the expressions for the internal friction and the sound velocity read [143]

$$Q^{-1} = \frac{\beta g^2}{\varrho v^2}\int_0^\infty\int_0^\infty d\Delta_0 d\Delta\, P(\Delta_0,\Delta)f(\Delta,E)\frac{\omega\tau_{\text{th}}}{1+\omega^2\tau_{\text{th}}^2},\qquad(7.104)$$

$$\frac{\delta v}{v} = -\frac{\beta g^2}{2\varrho v^2} \int_0^\infty \int_0^\infty \mathrm{d}\Delta_0 \mathrm{d}\Delta P(\Delta_0, \Delta) f(\Delta, E) \frac{1}{1+\omega^2 \tau_{\mathrm{th}}^2}, \tag{7.105}$$

where the ground-state properties appear in the factor

$$f(\Delta, E) = (\Delta^2/E^2) \cosh(\beta E/2)^{-2}, \tag{7.106}$$

with $E = \sqrt{\Delta_0^2 + \Delta^2}$.

One still has to average over the activation energy V. Starting from the WKB expression for the tunneling energy

$$\Delta_0 \approx \frac{2E_0}{\pi} e^{-\lambda}, \tag{7.107}$$

which with tunneling parameter $\lambda = (d/2\hbar)\sqrt{2mV}$ and zero-point energy E_0, Tielbürger et al. express V as a function of Δ_0 by inverting (7.107). To this purpose they assume the double-well potential to be given by a double oscillator with fixed single-well frequency $E_0/\hbar$, $V(x) = \frac{1}{2}m(E_0/\hbar)^2 x^2$. The barrier between the two wells then varies with their distance according to $V(0) = \frac{1}{8}m(E_0/\hbar)^2 d^2$. For the tunneling parameter one finds

$$\lambda = \frac{V}{E_0} \tag{7.108}$$

and easily deduces the uniform distribution

$$P(\Delta, V) = \frac{\overline{P}}{E_0}. \tag{7.109}$$

As a smooth physical cut-off for the barrier distribution, Tielbürger et al. propose a Gaussian law with width σ_0 [143],

$$P(\Delta, V) = \frac{\overline{P}}{E_0} \exp\left(-V^2/2\sigma_0^2\right). \tag{7.110}$$

For relaxational absorption at low temperatures, relatively small barriers are most significant, where (7.109) and (7.110) are equivalent. When we use the uniform distribution function (7.109) and take $\Delta \approx E \approx k_B T$, the integrals in (7.104) and (7.105) are easily evaluated and one finds [143]

$$Q^{-1} = \frac{\pi C k_B T}{E_0} \equiv \frac{\pi}{2} \frac{T}{T_a}, \tag{7.111}$$

$$\frac{\delta v}{v} = \frac{C k_B T}{E_0} \log(\omega \tau_0), \tag{7.112}$$

for $\omega \tau_0 \ll 1$. Thus, again, the absorption increases linearly, and the sound velocity decreases linearly above $T_a \approx 5$ K. When we compare these expressions with (7.101) and (7.103), we see that additional parameters E_0 and τ_0 have been introduced. From the relation $\omega \tau_{\mathrm{th}} = 1$, one expects the maximum in the absorption to occur at

$$k_B T_{\max} = -\frac{V_{\max}}{\log(\omega\tau_0)}, \tag{7.113}$$

i.e., the maximum temperature varies with frequency as $T_{\max} \propto 1/\log(\omega)$.

A word of caution is in order concerning the link between tunneling amplitude and activation energy, (7.107–7.110). At 5 K, only low barriers of the order of 50 K or smaller contribute to relaxation. Both the semiclassical expression (7.107) and the activated rate (7.85) are valid only if there are many vibrational levels below the top of the barrier (compare Fig. 6.6); for atomic tunneling, however, this condition can only be satisfied by potentials of at least several hundred K. In view of this argument, it seems doubtful whether thermally activated behavior above 5 K really reflects classical barrier crossing; hence (7.110) is rather to be considered as a phenomenological distribution function leading to the laws (7.111,7.112) (cf. [140]).

Finally we compare these findings with the results obtained from the soft-potential model. Owing to the slightly modified distribution function, the SPM [144] leads to $Q_{\mathrm{TS}}^{-1} \propto T^{3/4}$ and, accordingly, $(\delta v/v)_{\mathrm{TS}} \propto -T^{3/4}$. A linear contribution to $\delta v/v$ arises only from local low-frequency oscillators [146], which in fact closely resemble the low-barrier systems of Tielbürger et al.

7.5 Comparison with Experimental Results

Acoustic properties of amorphous solids have been a field of intense research for more than two decades [107,109]. Recently a survey of a large amount of data on glasses and disordered crystals was carried out by Pohl [156]; data and literature on polymers may be found in [157], and on polycrystalline metals in [153–155]. Although the tunneling model had originally been proposed for oxide glasses, it has been applied since then on various amorphous materials. The presence of low-energy tunneling states seems to be a general feature of configurational disorder; the amazing similarity observed in heat release, conductivity, and acoustic experiments indicates the universal aspect of the tunneling model [109,157,158].

Usually the break-down of the two-state description is related to the increase of internal friction at about 5 K. The following discussion serves two purposes. It renders evident the fact that the universal behavior is not restricted to temperatures below about 1 K, and it supports the theory that the two-level picture is valid up to about 20 K, yet with relaxation dynamics arising from incoherent tunneling. We present a few characteristic experiments on sound velocity and internal friction, which are closely related to the theory given above. (Since this theory is based on thermal equilibrium for the considered degrees of freedom, it totally misses the non-ergodic behavior that has been observed through the time-dependent specific heat at low temperatures.)

Oxide Glasses

Oxides of Si, Ge, and B form a particular amorphous state, which is sometimes characterized as an undercooled high-viscosity liquid [178]. When we compare low-temperature properties of this glass state with those of its crystalline counterpart, we find a fundamentally different behavior (see, e.g., Figs. 6.1 and 6.2). Here we discuss sound velocity and internal friction data for GeO_2 and B_2O_3.

We begin with the internal friction Q^{-1}. Figure 7.1 shows data for a-GeO_2 as a function of temperature, as measured with a vibrating-reed technique at a frequency of 6.3 kHz [128]. Below 100 mK the attenuation increases with temperature, corresponding to the range $\omega\tau_{\min} > 1$ or $T < \widetilde{T}$ of (7.96). (Yet the initial rise of the internal friction is not proportional to T^3 as predicted by (7.100) for $T < \widetilde{T}$.) Between 200 mK and 2 K the internal friction is constant, in agreement with (7.101) for $\widetilde{T} < T < T^*$. Above 3 K it increases up to a peak at about 100 K, and then falls off sharply.

The maximum temperature varies with the applied frequency ω; Fig. 7.2 shows data obtained in both acoustic [125,127] and dielectric experiments [126]; note the semilogarithmic presentation. Although they cover less than a decade in temperature, these data provide evidence that the thermally activated process prevails at the relaxation peak temperature, as is most obvious by comparing (7.102, 7.113) with Fig. 7.2.

The data in Fig. 7.1 have been fitted in two different ways, both relying on the parameters given in Table 7.5. The solid line in Fig. 7.1 follows

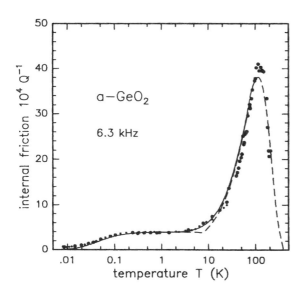

Fig. 7.1. Internal friction Q^{-1} for a-GeO_2 as a function of temperature. The dashed line accounts for thermal activated relaxation, the solid line for incoherent tunneling above 5 K. By courtesy of Rau et al. [128]

from the result from mode-coupling theory, (7.96), which accounts for the one-phonon process at low temperature $T < T^*$ and for incoherent tunneling above T^*. The curve was not continued beyond 40 K since at the absorption peak thermally activated relaxation is dominant. (Note, however, that with an appropriate lower cut-off for the parameter r, the mode-coupling result (7.96) accounts well for the existence of a relaxation peak at a given frequency [96]; a discrepancy arises only when comparing data taken at different frequencies.) We remark that a slight modification of the parameters would improve the fit below 200 mK, yet induce corresponding deviations above 5 K. The dashed line was obtained using (7.104) with the distribution (7.110), and including the one-phonon process at low temperatures. The additional fitting parameters are listed in Table 7.5 on p. 146.

Similar fits have been performed for SiO_2 [96], and for two samples of boroxide containing 130 ppm and 1.6 % OH, and labeled dry and wet B_2O_3, respectively [128]. These results differ so little from those shown in Fig. 7.1 that we refrain from reproducing them here.

In the intermediate temperature range from 4 to 20 K, two features in favor of the expression obtained from mode-coupling theory are to be mentioned. *First*, the solid curve provides a better fit between 4 and 20 K; the bump and the excessively steep increase in the dashed line cannot be removed by modification of the fitting procedure. *Second*, mode-coupling theory does not permit for additional parameters. The values for C, g, ρ, and v given in Table 7.5 are entirely fixed by the behavior of the internal friction below T^*; thus both the cross-over temperature T^* and the slope of the low-temperature

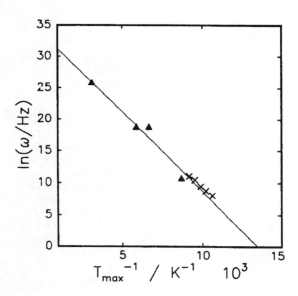

Fig. 7.2. Relation between the relaxation peak temperature T_{max} and the applied frequency ω. By courtesy of Rau and Hunklinger [127]

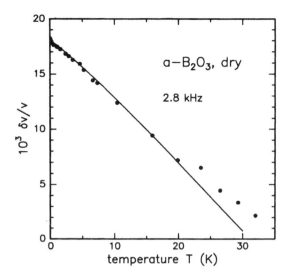

Fig. 7.3. Change of the sound velocity $\delta v/v$ of dry a-B_2O_3. The solid line is a fit according to (7.103). By courtesy of Rau et al. [128]

side of the relaxation peak are to be considered as a result of the theory obtained without adjustable parameters. The fit of the experimental data up to about 50 K shows a remarkably good agreement. (The thermally activated process fits the data noticeably less well, although there are free parameters to be adjusted.)

Now we turn to the relative change of sound velocity $\delta v/v$ as a function of T. At low temperatures, one observes logarithmic laws with both signs [107,110,114]; a typical example with a fit curve according to (7.103) for $T < T^*$ is provided by Fig. 6.2. Here we concentrate on the behavior above 1 K, where in many experiments a linear decrease with temperature has been found [128,135–137], in connection with a logarithmic frequency dependence of the prefactor (see Fig. 6.3). In Fig. 7.3 we compare data measured for dry B_2O_3 with (7.103) using the parameters given in Table 7.5. Up to 20 K,(7.103) fits well the observed linear temperature dependence. Again, similar results have been obtained for all samples for which the parameters are gathered in Table 7.5 [96,128]. A fit of comparable quality is also provided by (7.112) based on thermally activated relaxation (not shown in Fig. 7.3).

We stress that the solid curves in Fig. 7.3 and 7.1 are obtained with the numbers given in the table; thus both the slope of the sound velocity and the attenuation above 5 K are calculated using the parameters that have been determined from the low-temperature internal friction.

Table 7.1. Parameters of the fits for the data on oxide glasses. The deformation potential g and the sound velocity v are appropriately averaged over the longitudinal and transverse phonon branches, $g^2 \equiv g_\ell^2 = 2g_t^2$ and $v^{-5} = v_\ell^{-5} + v_t^{-5}$. The parameters E_0, τ_0, and σ_0 involve the activated process only.

	a-GeO$_2$	a-B$_2$O$_3$ (dry)	a-B$_2$O$_3$ (wet)	a-SiO$_2$
C	2.45×10^{-4}	3.8×10^{-4}	3.6×10^{-4}	2.8×10^{-4}
g / eV	1.35	0.65	0.65	2.2
ρ / (gcm^{-3})	3.6	1.8	1.8	2.2
v / (ms^{-1})	2814	2310	2373	2040
E_0 / K	15	15	15	
τ_0 / s	1.6×10^{-13}	10^{-13}	10^{-13}	
σ_0 / K	2200			

Polycrystalline Metals

Recently Esquinazi and his co-workers reported most surprising acoustic measurements on various polycrystalline metals (Ag, Al, Cu, Nb, Pd, Pt, and Ta) [152–155], showing low-temperature properties very similar to those of oxide glasses. The variation of sound velocity and attenuation with temperature, frequency, and applied strain was found to be quantitatively comparable to that observed for glasses. As an example, we show in Fig. 7.4 sound velocity data for polycrystalline Al both in the superconducting and in the normal state. Note the similarity to the data for vitreous silica in Fig. 6.2.

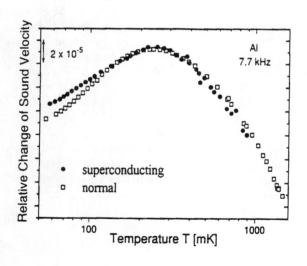

Fig. 7.4. Temperature dependence of the sound velocity for polycrystalline Al at 7.7 kHz. The normal state is achieved by applying a magnetic field $B = 10.7$ mT and a strain $\epsilon = 9 \times 10^{-5}$; for the superconducting state these parameters read $B = 4.7$ mT and $\epsilon = 7 \times 10^{-5}$ By courtesy of König et al. [154]

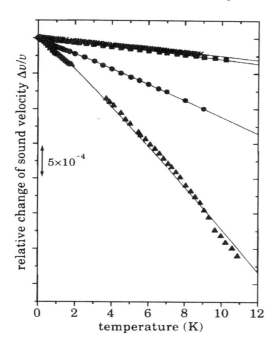

Fig. 7.5. Change of sound velocity $\delta v/v$ with temperature for polycrystalline Pt (crosses), PtRh (squares), Pd (circles), and Ta (triangles). For fitting procedure and parameters see text. By courtesy of Gaganidze et al. [155]

If we assume that the anomalous temperature dependence arises from two-level systems, we would expect a behavior similar to that observed for metallic glasses [133]. According to Fig. 7.4, polycrystalline metals behave rather like insulating glasses. This has led to the conclusion that the two-level systems present in the former do not interact with the conduction electrons. This conjecture is confirmed by data obtained for the normal and superconducting state for a given Al sample, which show the sound velocity to be almost independent of the state of the conduction electrons (see Fig. 7.4).

At temperatures above 1 K, the sound velocity decreases linearly with temperature. Figure 7.5 shows data for polycrystalline Pt, PtRh, Pd, and Ta, at frequencies between 420 Hz and 20 kHz and temperatures up to 11 K. The solid lines represent fits according to (7.103) for $T > T^*$ performed by the authors of [155], where C has been deduced from the logarithmic law observed at very low temperatures, and where the cross-over temperature T^* has been treated as a free parameter. The values used for the fits in Fig. 7.5 read 5.7 K, 10 K, 5.6 K, and 3.7 K for Pt, PtRh, Pd, and Ta. Obviously, these values contradict the assumption $T > T^*$; if one suspects some uncertainty in the determination of the parameter C, one might well obtain a sufficiently low cross-over temperature as to permit the use of the linear law in (7.103).

Clearly, the most intriguing question is that of the origin of the anomalous behavior and the microscopic nature of the two-level systems. The mystery is all the more complete as the authors of [155] state that the two-level systems cannot be ascribed only to the grain boundaries of the polycrystalline material.

Polymers

Figure 7.6 displays internal friction data for polystyrene (PS) obtained by Nittke et al. [157] at two frequencies 240 Hz and 3.2 kHz over three decades with respect to temperature. Except at the bounds of the temperature range considered, there is a striking resemblance to data for oxide glasses (see, e.g., Fig. 7.1). At the low-temperature side, the plateau edge has not yet been reached at 100 mK; for PMMA Federle and Hunklinger [159] observed a falling off just below 1 K at much higher frequency, in accordance with (7.99). The attenuation increase after the relaxation peak at about 80 K indicates an additional dissipation mechanism, which dominates at high temperatures.

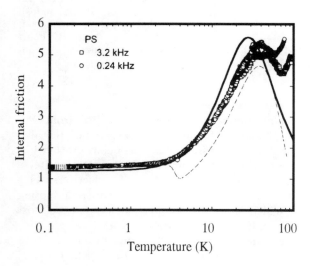

Fig. 7.6. Internal friction of polystyrene as a function of temperature at two different frequencies. The dashed line accounts for thermally activated relaxation and the solid one for incoherent tunneling above 5 K, at a frequency of 240 Hz. By courtesy of Nittke et al. [157]

The data in Fig. 7.6 are compared with the two different models already used for oxide glasses. The dashed line represents a fit by Nittke et al. [157] based on the one-phonon process arising from the tunneling model and a thermally activated process according to (7.104); note the similarity to the dashed line in Fig. 7.1. The solid line is given by (7.100) for $T > \widetilde{T}$. The values for the respective fit parameters read $C = 8.3 \times 10^{-4}$, $\omega/2\pi = 0.24$ kHz, $\tau_0 = 10^{-17}$ s, $E_0 = 13$ K, $\sigma_0 = 1200$ K, and $T^* = 3$ K. As for oxide glasses, the activated process provides quite a poor fit at the low-temperature side of the relaxation peak, in spite of three adjustable parameters, τ_0, E_0, and σ_0.

7.6 Summary

In the previous section we discussed low-temperature sound propagation in oxide glasses, polycrystalline metals, and polymers. In these materials, both sound velocity and internal friction exhibit a characteristic temperature dependence which below 1 K is well accounted for by the tunneling model with the one-phonon relaxation process. Well above 1 K the sound velocity decreases linearly with temperature, whereas the attenuation increases towards a relaxation peak at about 50 ... 100 K.

There is a general agreement about the low-temperature acoustic properties being determined by the interaction with two-level tunneling systems. The origin of the behavior above 3 K is not yet settled. From the comparison of two theoretical models with experiments, we are led to distinguish four regimes with respect to the dynamics of the tunneling systems.

(i) Resonant interaction. At low temperatures $T < 5$ K, the TSs show weakly damped oscillations. At moderate frequencies, a relevant absorption process for sound waves is given by the interaction with symmetric TS, where the energy splitting fulfils the resonance condition $E = \hbar\omega$, (7.91). At higher temperatures or low frequencies, $k_B T \gg \hbar\omega$, this process is suppressed, since both levels are equally occupied. Regarding the sound velocity, the resonant interaction results in a logarithmic increase with temperature, (7.103).

(ii) Relaxation of asymmetric TSs. At temperatures below 5K, sound attenuation via relaxational absorption occurs because there are TSs satisfying the condition $\omega\tau_1 \approx 1$. Owing to the weak-damping condition $\hbar\widetilde{\Gamma} \ll E$, only asymmetric TSs contribute with an amplitude Δ^2/E^2. As a consequence, one finds a plateau in the internal friction and the logarithmic decrease of the sound velocity.

(iii) Relaxation of overdamped TSs. According to our discussion in Sect. 7.2, the tunneling oscillations disappear at $k_B T^* = 2/\pi\widetilde{\gamma}$. For $T > T^*$ the motion of the TSs between the two wells occurs by incoherent tunneling; accordingly the correlation function is that of a Debye relaxator with rate $1/\tau_1$ [cf. (7.71, 7.77)]. Inserting measured values for the parameters yields $T^* \approx 5$ K. We find a linear increase with temperature for the internal friction, and a linear decrease for the sound velocity.

(iv) Thermally activated relaxation. At even higher temperatures, the two-level description becomes invalid. The finite occupation of excited states in the double-well potential results in thermally activated relaxation with a rate that can be described by an Arrhenius law $\tau_{\text{th}}^{-1} = \tau_0^{-1} e^{-V/k_B T}$. For τ_{th}^{-1} much larger than the quantum tunneling rate $\widetilde{\Gamma}$, quantum effects have disappeared and the motion in the double-well potential of Fig. 1 can be described by classical mechanics. Because of its dependence on the unknown parameters τ_0 and V, the cross-over temperature has to be inferred from experiment. In view of the available data, we would situate this cross-over above 20 K.

For $T < 5$ K relaxation of tunneling systems with finite asymmetry and $\omega\tau_1 \approx 1$ appears to be the most effective dissipation mechanism. The relaxation process of biased TSs has been elaborated by Jäckle [105] using Fermi's Golden Rule; in the weak-damping limit $\hbar\widetilde{\Gamma} \ll E$, corresponding to ranges (i) and (ii), our perturbative approach yields results identical to Jäckle's.

At about 5 K, experiments show a significant change in the dynamic behavior of tunneling systems, which the usual description based on two-state approximation and perturbation theory fails to account for; yet it is not clear from the beginning which feature has to be modified in order to describe properly the observed relaxation dynamics. Predominance of a thermally activated rate implies the break-down of the two-level picture [cf. (iv)], whereas incoherent tunneling according to (iii) still relies on the two-state approximation but requires us to go beyond perturbation theory and to treat the interaction of the tunneling system with phonons in a strong-coupling approach.

The mode-coupling approach yields a qualitative change in the dynamics at a temperature $T^* = 2/k_B\pi\widetilde{\gamma}$; this picture is supported by higher-order terms of the perturbation series, which indicate the perturbative approach to be invalid for $T > T^*$. This mode-coupling theory does not allow for additional parameters; the onset temperature T^* and the slopes of internal friction and sound velocity are determined by the coupling constant $\widetilde{\gamma}$. Thus we are led to ascribe the temperature dependence at $T \geq 5$ K to overdamped two-level systems.

Adjusting the only parameters $\widetilde{\gamma}$ and C, we find that the theory agrees surprisingly well with various experimental quantities: the low-temperature increase of the internal friction; its plateau value; the onset temperature of the increase towards the relaxation peak and the corresponding slope; the slope of the linear decrease of sound velocity between 1 and about 20 K. Both the strong increase in the internal friction and the linear temperature dependence of sound velocity observed above 5 K can be derived from both incoherent tunneling and thermal activation; yet the mode-coupling theory does not allow for additional parameters and fits the data better than the thermally activated process.

Owing to its much stronger temperature dependence, thermal activation will at some point exceed the quantum tunneling rate, ultimately leading to a classical Arrhenius behavior. The frequency dependence of the maximum of the relaxation peak, $T_{\max} \propto \log(\omega)$ clearly indicates an activated rate; from this we conclude that at 40 K, most systems are already in the thermally activated regime. (If we neglected thermal activation, the mode-coupling approach would yield a variation of the peak temperature with $\sqrt{\omega}$.)

There remain discrepancies at very low temperatures for both sound velocity and attenuation. The measured internal friction does not increase as $Q^{-1} \propto T^3$ as expected from Jäckle's theory [105]; one finds instead a power law with an exponent between one and two. Similar results were reported earlier for coverglass and Suprasil W [114,152]. Burin and Kagan proposed an explanation in terms of a resonant interaction of two-level systems [148,112]. As to the sound velocity, the prefactors of the logarithmic laws below and above $\widetilde{T}$ do not obey the ratio 2:−1 as expected from (7.103); instead most experiments show a ratio of 1:−1.

In summary, we find several ranges for the dynamics of tunneling systems in glasses. Below 5 K, the weakly damped coherent oscillations are well described by perturbation theory; as the most prominent features, we note the logarithmic temperature dependence of sound velocity and the plateau of the internal friction. Above 5 K the thermal motion of the atoms destroys the phase coherence of the two-level systems, thus requiring a strong-coupling theory; our mode-coupling approach provides a quantitatively correct description of the temperature dependence. Above 20 K the two-state approximation ceases to be valid; thermal occupation of excited levels results in an activated rate and ultimately permits a description in terms of classical mechanics.

8. Small-Polaron Approach

8.1 Break-Down of Perturbation Theory

When calculating the memory matrix (7.26) in lowest-order perturbation theory, we have retained the dissipative part only, and discarded the reactive one. Accordingly, the phonon coupling affects the tunneling dynamics only through the damping rates, which involve low-frequency lattice modes. In this chapter we include high-frequency phonons and we discuss how they bear upon the tunneling motion. To this purpose we consider the Hamiltonian

$$H = \frac{1}{2}\Delta_\mathrm{b}\sigma_x + \frac{1}{2}\sigma_z \sum_k \hbar\lambda_k(b_k + b_k^\dagger) + \sum_k \hbar\omega_k b_k^\dagger b_k, \tag{8.1}$$

which corresponds to (6.16) with zero asymmetry, $\Delta = 0$, and Δ_0 replaced by Δ_b. Here Δ_b denotes the bare tunneling amplitude.

For vanishing phonon coupling, the eigenvalues of the spin part read $\pm\frac{1}{2}\Delta_\mathrm{b}$. When we assume $\Delta_\mathrm{b} \ll k_\mathrm{B}T \ll \hbar\omega_\mathrm{D}$, second-order perturbation theory in terms of λ_k yields with (8.6,8.7) a finite energy shift

$$\Delta E = -\Delta_\mathrm{b}\left(\frac{1}{2}\frac{\alpha}{\tilde{\omega}^2}\omega_\mathrm{D}^2 + \frac{1}{3}\pi^2\frac{\alpha}{\hbar^2\tilde{\omega}^2}k_\mathrm{B}^2 T^2\right). \tag{8.2}$$

This correction comprises a constant and a temperature-dependent part; the former results from the coupling to high-frequency phonons, and the latter from thermal phonons. The minus sign corresponds to a slowing down of the tunneling motion.

Yet for parameters describing real systems, the constant term in brackets is larger than unity in general. Hence the perturbation series does not converge; nevertheless we may conclude from the second-order term that high-frequency phonons lead to a significant reduction of the zero-temperature tunnel energy. Since it is purely static, this effect may be absorbed in an effective tunnel energy Δ_0.

The second term in brackets in (8.2) strongly increases with temperature. Its approaching unity signifies the break-down of finite-order perturbation theory. This picture is supported by the perturbation series for the damping rate, which in terms of the second-order rate γ_2 reads as

154   8. Small-Polaron Approach

$\gamma_2[1 + \frac{2}{3}\pi^2\alpha(k_{\rm B}T/\hbar\tilde\omega)^2 + ...]$. Comparison with (7.82) reveals that this breakdown occurs precisely at the temperature of cross-over to incoherent motion, as obtained from the mode-coupling theory in the previous chapter.

Yet the mode-coupling approach accounts for only the enhancement of the rate; the effect of the heat bath on the bare tunnel energy has been discarded. From the above observations it is clear that a proper treatment of the influence of high-frequency phonons requires a partial resummation of the infinite perturbation series, the first term of which is given by (8.2).

In this chapter we present a different approach to the two-state dynamics, which is based on the well-known polaron transformation and which has been widely used in works on the spin-boson problem. It permits us, in particular, to circumvent the problems arising from the divergence of the perturbation series stated above.

8.2 Small-Polaron Representation

Applying the canonical transformation

$$S = \exp\left(-\sigma_z \sum_k \frac{\lambda_k}{\omega_k}(b_k - b_k^\dagger)\right) \tag{8.3}$$

on the Hamiltonian (8.1) yields, with $H \to e^S H e^{-S}$,

$$H = \frac{\Delta_{\rm b}}{2}(\sigma_+ B_- + \sigma_- B_+) + \sum_k \hbar\omega_k b_k^\dagger b_k, \tag{8.4}$$

where we have neglected a constant and used

$$B_\pm = \exp\left(\pm \sum_k \frac{\lambda_k}{\omega_k}(b_k - b_k^\dagger)\right) \tag{8.5}$$

and the ladder operators $\sigma_+ = |L\rangle\langle R|$ and $\sigma_- = |R\rangle\langle L|$. (Note $\sigma_x = \sigma_+ + \sigma_-$.) Such a form is well known from studies on polaron motion [86] and dissipative two-level dynamics [15–25].

The properties of the oscillator heat bath are concisely described by means of the spectral function

$$J(\omega) = \frac{\pi}{2}\sum_k \lambda_k^2 \delta(\omega - \omega_k). \tag{8.6}$$

For lattice vibrations of a three-dimensional solid, one finds the cubic spectral density

$$J(\omega) = \pi(\alpha/\tilde\omega^2)\omega^3 \equiv \pi(\hbar^2/k_{\rm B}^2)\tilde\alpha\omega^3, \tag{8.7}$$

with the dimensionless coupling constant α; only the quantity $(\alpha/\tilde\omega^2)$ is of physical significance. The cubic law arises from the product of the Debye density of states and the frequency-dependent scattering matrix element. Strictly speaking it is valid only in the limit of large wave-lengths, i.e., well below the Debye frequency; since the precise form of the high-frequency cut-off is of little importance, we use the cubic law up to the Debye frequency ω_D and use a sharp cut-off function. In this chapter we will use the constant $\tilde{\alpha}$ whose dimension is (temperature)$^{-2}$.

Whereas the Ohmic damping model has attracted much attention during the last decade [9–25], only a few studies have been carried out on the superohmic heat bath; some of the available results are contradictory. Pirc and Gosar treated the first term in (8.4) as a perturbation and found weakly damped tunneling oscillations at low temperatures and relaxation at high temperatures. From a real-time path integral approach it has been concluded that the cubic spectral density would not lead to incoherent motion [15]. Considering overdamped motion arising from a heat bath with both linear and cubic terms, Grabert found a cross-over from the Ohmic rate that varies with inverse temperature to a phonon-induced relaxation rate that increases exponentially with temperature [91]. A mode-coupling approach yielded a cross-over from coherent oscillations at low temperatures to incoherent motion; the high-temperature rate was found to decrease with rising temperature [95,128].

Here we treat the dynamics arising from (8.1)–(8.7) in a small-polaron representation which is closely related to Holstein's work on diffusion of a dressed particle in a crystal [86]. First we derive the relevant correlation function, whose dissipation kernel is given as a power series in terms of the tunneling amplitude. After we consider the undamped case in Sect. 8.3, we evaluate this kernel in Sects. 8.4–8.7 by means of various approximation schemes. At low temperatures the spin-phonon coupling constitutes a weak perturbation; a corresponding power series expansion is shown to yield the correct weak-damping behavior and to diverge at some critical temperature T_0. For the high-temperature limit $T \gg T_0$ we find exact upper and lower bounds for the damping term, which are compared with the well-known saddle-point approximation. In the last section of this chapter the results are discussed and related to previous work.

Equation of Motion and Correlation Function

All relevant dynamic information is contained in the two-time correlation function of the position operator σ_z,

$$G(t - t') = \langle \sigma_z(t)\sigma_z(t') \rangle, \tag{8.8}$$

where the angular brackets denote the average with respect to the density operator ρ, $\langle ... \rangle = \text{tr}(\rho ...)/\text{tr}(\rho)$.

Since σ_z is invariant under the canonical transformation, i.e., $[S, \sigma_z] = 0$, the correlation function (8.8) can be calculated with either Hamiltonian (8.1) or (8.4); throughout this chapter we will use the latter.

Time evolution of any operator A is governed by von Neumann's equation $\hbar \dot{A} = i[H, A]$; for the pseudo-spins we easily find with (8.4)

$$\hbar \dot{\sigma}_z = i\Delta_b (B_+ \sigma_- - B_- \sigma_+), \tag{8.9}$$

$$\hbar \dot{\sigma}_\pm = \mp \frac{i}{2} \Delta_b B_\pm \sigma_z. \tag{8.10}$$

When we insert the formal integral of the second equation,

$$\sigma_\pm(t) = \sigma_\pm(t') \mp \frac{i}{2\hbar} \Delta_b \int_{t'}^{t} d\tau B_\pm(\tau) \sigma_z(\tau), \tag{8.11}$$

in the first one and multiply on the right-hand side with $\sigma_z(t')$, we find an integro-differential equation for $\sigma_z(t)\sigma_z(t')$,

$$\frac{\partial}{\partial t} \sigma_z(t)\sigma_z(t') + \int_{t'}^{t} d\tau K(t,\tau) \sigma_z(\tau)\sigma_z(t') + I(t,t') = 0, \tag{8.12}$$

where the inhomogeneity reads with $\sigma_\pm(t')\sigma_z(t') = \mp \sigma_\pm(t')$

$$I(t,t') = \frac{i}{\hbar} \Delta_b (B_-(t)\sigma_+(t') + B_+(t)\sigma_-(t')), \tag{8.13}$$

and the kernel is defined as

$$K(t,t') = \frac{1}{2\hbar^2} \Delta_b^2 (B_+(t)B_-(t') + B_-(t)B_+(t')). \tag{8.14}$$

Now we integrate (8.12) from t' to t, insert the left-hand side repeatedly in the right-hand side, and take the thermal average. As a result, an infinite series for the correlation function (8.8) is engendered,

$$\begin{aligned} G(t-t') &= 1 - \int_{t'}^{t} d\tau_1 \int_{t'}^{\tau_1} d\tau_2 \langle K(\tau_1,\tau_2) \rangle \\ &+ \int_{t'}^{t} d\tau_1 \int_{t'}^{\tau_1} d\tau_2 \int_{t'}^{\tau_2} d\tau_3 \int_{t'}^{\tau_3} d\tau_4 \langle K(\tau_1,\tau_2)K(\tau_3,\tau_4) \rangle + \ldots \\ &- \int_{t'}^{t} d\tau \langle I(\tau,t') \rangle \\ &+ \int_{t'}^{t} d\tau_1 \int_{t'}^{\tau_1} d\tau_2 \int_{t'}^{\tau_2} d\tau \langle K(\tau_1,\tau_2)I(\tau,t') \rangle + \ldots . \end{aligned} \tag{8.15}$$

(Since the thermal average is invariant under time translation, the correlation function depends only on the difference of its arguments.) In many cases it proves convenient to consider the Laplace transform

$$G(z) = i \int_0^\infty dt e^{izt} G(t), \tag{8.16}$$

which has an imaginary part that yields the correlation spectrum

$$G''(\omega) = \operatorname{Im} \lim_{\epsilon \to 0} G(\omega + i\epsilon). \tag{8.17}$$

The coupling constant of a single bath mode varies as $\lambda_k \propto N^{-1/2}$ with the number of particles N. Thus in the thermodynamic limit $N \to \infty$, the interaction of each bath mode is infinitely weak. Regarding the spin dynamics, the sum of all bath degrees of freedom results in finite quantities. Yet in the time evolution of the operators (8.5), the spin part of (8.4) may be neglected,

$$B_\pm(t) = \mathrm{e}^{\mathrm{i}H_\mathrm{B}t/\hbar} B_\pm \mathrm{e}^{-\mathrm{i}H_\mathrm{B}t/\hbar}, \tag{8.18}$$

with $H_\mathrm{B} = \sum_k \hbar\omega_k b_k^\dagger b_k$.

In order to obtain explicit results from the formal expressions (8.15, 8.16), we are required to use an approximative treatment with respect to the time fluctuations of the operators $B_\pm$. It proves to be convenient to separate the average value $\langle B_\pm \rangle$ and the fluctuating part

$$\xi_\pm(t) = B_\pm(t) - \langle B_\pm \rangle. \tag{8.19}$$

8.3 Limit of Zero Damping

First we consider the undamped case. In terms of (8.19), this corresponds to discarding the fluctuations $\xi_\pm$. The average

$$\tilde{\Delta}_0 = \Delta_\mathrm{b} \langle B_\pm \rangle \tag{8.20}$$

defines an effective tunneling amplitude that is reduced from its bare value by a phonon overlap matrix element. With $\xi_\pm \equiv 0$, neither the kernel $K = \Sigma_0$ nor the inhomogeneity I_0 depend on time,

$$\Sigma_0 = \tilde{\Delta}_0^2/\hbar^2, \tag{8.21}$$

$$I_0 = \mathrm{i}(\tilde{\Delta}_0/\hbar)\sigma_x; \tag{8.22}$$

when we take $\xi_\pm \equiv 0$ in (8.4), the statistical operator factorizes as

$$\rho_0 = \frac{1}{Z}\Big(1 - \sigma_x \tanh(\beta\tilde{\Delta}_0/2)\Big)\mathrm{e}^{-\beta H_\mathrm{B}}. \tag{8.23}$$

After we insert (8.21–8.23) in the series (8.15) and perform the time integrations, we obtain a power series which can be rearranged to yield sine and cosine functions of argument $\tilde{\Delta}_0(t-t')/\hbar$,

$$G_0(t-t') = \cos[\tilde{\Delta}_0(t-t')/\hbar] - \mathrm{i}\tanh(\beta\tilde{\Delta}_0/2)\sin[\tilde{\Delta}_0(t-t')/\hbar]. \tag{8.24}$$

The first term of the right-hand side gives the symmetrized correlation function, and the second term the commutator response function.

With $I_0(z) = -\mathrm{i}(\tilde{\Delta}_0/\hbar z)\sigma_x$, the Laplace transform reads

$$G_0(z) = \frac{z + \tilde{\Delta}_0 \tanh(\beta\tilde{\Delta}_0/2)}{z^2 - \tilde{\Delta}_0^2}. \tag{8.25}$$

Upon Fourier transformation, or using (8.17), we find the correlation spectrum to consist of two delta peaks at $\pm\tilde{\Delta}_0$,

158   8. Small-Polaron Approach

$$G_0''(\omega) = \frac{\pi}{2} \sum_{\pm} [1 \pm \tanh(\beta\tilde{\Delta}_0/2)]\delta(\omega \mp \tilde{\Delta}_0), \qquad (8.26)$$

which have amplitudes that account for the thermal occupation of the upper and lower energy eigenstates of a two-level system. The terms proportional to $\tanh(\beta\tilde{\Delta}_0/2)$ arise from the inhomogeneity I; they are negligible for sufficiently high temperatures $T \gg \tilde{\Delta}_0/k_B$.

8.4 Approximation Scheme

The treatment of the dissipative two-state dynamics in this chapter is based on two approximations. First, when we assume temperatures well above the two-level splitting, $\tilde{\Delta}_0 \ll k_B T$, we neglect the inhomogeneity I_0 in (8.15). Second, we expand the damping function in terms of the fluctuation operators (8.19).

Initial State

We assume the spin part of the Hamiltonian to be small compared with the temperature, which yields a particular form of the initial state,

$$\rho = e^{-\beta H_B}, \qquad (8.27)$$

with $H_B = \sum_k \hbar\omega_k b_k^\dagger b_k$. Since ρ does not depend on spin operators, any expectation value involving $\sigma_\pm$ or σ_z vanishes. From (8.18, 8.27) it follows that all terms in (8.15) involving the inhomogeneity I vanish, resulting in

$$\begin{aligned} G(t-t') &= 1 - \int_{t'}^{t} d\tau_1 \int_{t'}^{\tau_1} d\tau_2 \langle K(\tau_1, \tau_2) \rangle \\ &+ \int_{t'}^{t} d\tau_1 \int_{t'}^{\tau_1} d\tau_2 \int_{t'}^{\tau_2} d\tau_3 \int_{t'}^{\tau_3} d\tau_4 \langle K(\tau_1, \tau_2) K(\tau_3, \tau_4) \rangle + \ldots \end{aligned} \qquad (8.28)$$

For $\tilde{\Delta}_0 \ll k_B T$ the lowest-order term I_0 in (8.9) is in fact small. In the weak-damping limit, an expansion of the statistical operator in terms of the fluctuations ξ permits us to calculate next-order corrections and to conclude that the inhomogeneity I is insignificant. Yet this is less obvious for the (most interesting) case of strong damping. In the literature on the spin-boson model, this problem is often avoided by considering a particular initial state localized in one of the wells, e.g., $\sigma_z = 1$.

In the remainder of this chapter, we derive an approximation scheme for the correlation function (8.28) which may be viewed as a cumulant expansion in terms of the kernel $K(z)$ and which permits us to write the Laplace transform as

$$G(z) = -\frac{1}{z + \Sigma_0(z) + \Sigma_1(z) + \Sigma_2(z) + \ldots}. \qquad (8.29)$$

The first term is given by the zero-damping limit with

$$\Sigma_0(z) = -\frac{\tilde{\Delta}_0^2}{\hbar^2}\frac{1}{z}. \tag{8.30}$$

First-Order Approximation

We are going to replace the kernel $K(t,t')$ in (8.15) by its thermal average $\langle K(t,t')\rangle$. The operators $\sigma_+ B_-$ and $\sigma_- B_+$ in the Hamiltonian cause spin flips between the two states $\sigma_z = \pm 1$; accordingly the kernel $K(\tau,\tau')$ corresponds to a pair of flips at times τ and τ'. When we consider a term on the right-hand side of (8.15),

$$\langle ...K(\tau_1,\tau_2)...K(\tau_3,\tau_4)...\rangle, \tag{8.31}$$

we retain only correlations arising from a given K; accordingly, (8.31) factorizes as

$$\langle ...\rangle\langle K(\tau_1-\tau_2)\rangle\langle ...\rangle\langle K(\tau_3-\tau_4)\rangle\langle ...\rangle. \tag{8.32}$$

(Note that the thermal average $\langle K(t-t')\rangle$ depends only on the difference of the two time arguments of $K(t,t')$.)

Inserting (8.32) in the series for $G(t-t')$, we find the first-order approximation

$$\begin{aligned} G_1(t-t') &= 1 - \int_{t'}^{t} d\tau_1 \int_{t'}^{\tau_1} d\tau_2 \langle K(\tau_1-\tau_2)\rangle \\ &+ \int_{t'}^{t} d\tau_1 \int_{t'}^{\tau_1} d\tau_2 \int_{t'}^{\tau_2} d\tau_3 \int_{t'}^{\tau_3} d\tau_4 \langle K(\tau_1-\tau_2)\rangle\langle K(\tau_3-\tau_4)\rangle + ... \end{aligned} \tag{8.33}$$

This series may be rewritten as an integral equation,

$$G_1(t-t') = 1 - \int_{t'}^{t} d\tau \int_{t'}^{\tau} d\tau' \langle K(\tau-\tau')\rangle G_1(\tau'-t'). \tag{8.34}$$

The static part of $\langle K(t-t')\rangle$ is given by (8.21), $\lim_{t\to\infty}\langle K(t)\rangle = \tilde{\Delta}_0^2/\hbar^2$. The remaining two-time correlation function of the fluctuations (8.19) reads

$$\Sigma_1(\tau_1-\tau_2) = \frac{1}{2}\frac{\Delta_b^2}{\hbar^2}\sum_{\pm}\langle \xi_\pm(\tau_1)\xi_\mp(\tau_2)\rangle \tag{8.35}$$

and permits us to write the Laplace transform of (8.34) as

$$G_1(z) = \frac{-1}{z+\Sigma_0(z)+\Sigma_1(z)}. \tag{8.36}$$

Comparison with the expression derived from a path-integral approach reveal that (8.35, 8.36) are essentially equivalent to the *non-interacting blip approximation* [15,91], where a *blip* corresponds to the particle that is scattered once to and fro between the two wells.

Second-Order Approximation

Now we take into account correlations of two different factors K in the series (8.28). When we define

$$\delta K(\tau, \tau') = K(\tau, \tau') - \langle K(\tau - \tau')\rangle, \tag{8.37}$$

the difference of (8.31) and (8.32) reads to second order in δK

$$\langle ...\delta K(\tau_1, \tau_2)...\delta K(\tau_3, \tau_4)...\rangle. \tag{8.38}$$

As in (8.32) the time intervals $t-\tau_1$, $\tau_2-\tau_3$, and τ_4-t' may involve any number of factors $\langle K \rangle$. By taking the sum of such contributions and integrating over all additional times, we obtain

$$- G_1(t-\tau_1)G_1(\tau_2 - \tau_3)\langle \delta K(\tau_1, \tau_2)\delta K(\tau_3, \tau_4)\rangle G_1(\tau_4 - t'). \tag{8.39}$$

We still have to integrate over τ_1, τ_2, τ_3, and τ_4; after changing the integrations according to

$$\int_{t'}^{t} d\tau_1 \int_{t'}^{\tau_1} d\tau_2 \int_{t'}^{\tau_2} d\tau_3 \int_{t'}^{\tau_3} d\tau_4 ... = \int_{t'}^{t} d\tau_1 \int_{t'}^{\tau_1} d\tau_4 \int_{\tau_4}^{\tau_1} d\tau_2 \int_{\tau_4}^{\tau_2} d\tau_3 ... \tag{8.40}$$

and defining

$$\Sigma_2(t-t') = -\int_{t'}^{t} d\tau \int_{t'}^{\tau} d\tau' G_1(\tau - \tau')\langle \delta K(t, \tau)\delta K(\tau', t')\rangle, \tag{8.41}$$

we obtain the second-order term on the right-hand side of (8.28)

$$-\int_{t'}^{t} d\tau \int_{t'}^{\tau} d\tau' G_1(t-\tau)\Sigma_2(\tau - \tau')G_1(\tau' - t'). \tag{8.42}$$

As to the contribution of fourth order in δK,

$$\langle ...\delta K(\tau_1, \tau_2)...\delta K(\tau_3, \tau_4)...\delta K(\tau_5, \tau_6)...\delta K(\tau_7, \tau_8)...\rangle, \tag{8.43}$$

we retain only correlations of the first and the second factor, and of the third and the fourth. With (8.41) we thus find

$$\int_{t'}^{t} d\tau_1 \int_{t'}^{\tau_1} d\tau_2 \int_{t'}^{\tau_2} d\tau_3 \int_{t'}^{\tau_3} d\tau_4 G_1(t-\tau_1)\Sigma_2(\tau_1 - \tau_2) \\ \times \ G_1(\tau_2 - \tau_3)\Sigma_2(\tau_3 - \tau_4)G_1(\tau_4 - t'). \tag{8.44}$$

When we proceed in the same way for terms containing six, eight, etc. factors δK, we obtain an infinite series which reads schematically

$$G_2 = G_1 - G_1\Sigma_2 G_1 + G_1\Sigma_2 G_1 \Sigma_2 G_1 - G_1\Sigma_2 G_1 \Sigma_2 G_1 \Sigma_2 G_1 + ... \tag{8.45}$$

The first term is given by (8.34), the second one by (8.42) and the third one by (8.44). The series leads to the integral equation for G_2

$$G_2(t-t') = G_1(t-t') \\ - \int_{t'}^{t} d\tau \int_{t'}^{\tau} d\tau' G_1(t-\tau)\Sigma_2(\tau - \tau')G_2(\tau' - t'). \tag{8.46}$$

8.4 Approximation Scheme

In terms of a perturbation theory with respect to δK, this corresponds to retaining only the usual second-order contribution.

Upon performing the Laplace transformation we obtain

$$G_2(z) = G_1(z) - G_1(z)\Sigma_2(z)G_2(z), \tag{8.47}$$

and finally, after solving for G_2 and inserting (8.36),

$$G_2(z) = \frac{-1}{z + \Sigma_0(z) + \Sigma_1(z) + \Sigma_2(z)}. \tag{8.48}$$

Terms of Higher Order

Now we consider the terms involving three factors δK. When proceeding as above for Σ_2, we find the third-order contribution

$$\begin{aligned}\Sigma_3(t-t') &= \int_{t'}^{t}\mathrm{d}\tau \int_{t'}^{\tau}\mathrm{d}\tau' \int_{t'}^{\tau'}\mathrm{d}\tau'' \int_{t'}^{\tau''}\mathrm{d}\tau''' G_1(\tau-\tau') \\ &\quad \times\, G_1(\tau''-\tau''')\langle \delta K(t,\tau)\delta K(\tau',\tau'')\delta K(\tau''',t')\rangle.\end{aligned} \tag{8.49}$$

The contribution of fourth order is rather more complicated, since one has to subtract reducible terms involving twice Σ_1; since δK is not a simple field operator but an exponential of b_k and $b_k^\dagger$, the fourth-order contribution contains additional terms besides the usual crossing and rainbow diagrams.

Here we stop the expansion of the self-energy of (8.29). The nth approximation Σ_n is constructed by calculating the expectation value of n factors K, inserting $(n+1)$ factors G_{n-1}, and subtracting all reducible terms with respect to $\Sigma_0, ..., \Sigma_{n-1}$. When we note (8.14) and (E13), the resulting correlations of δK can be expressed in terms of the phase φ, (8.54).

Formally, the subsequent approximations $G_1, G_2, ...$ can be viewed as a non-crossing cumulant expansion of the self-energy in terms of $\delta K(t,t')$. In the next two sections we evaluate the first-order term $\langle \delta K(t,t')\rangle$ in detail, and the second-order term for two limiting cases.

Thermal Average

Since only harmonic oscillators are involved, any correlation function of the phonon operators (8.19) can be evaluated. In Appendix E we have gathered those which are relevant here.

Regarding the reduced tunneling amplitude, it will prove convenient to separate the temperature dependent part of the exponential factor from the remainder,

$$\Delta_0 = \Delta_{\text{b}} e^{-\frac{1}{2}\Sigma_k u_k^2}, \tag{8.50}$$

$$\tilde{\Delta}_0 = \Delta_0 e^{-\Sigma_k u_k^2 n_k}, \tag{8.51}$$

where we have used (E5) and

$$u_k = \lambda_k/\omega_k, \tag{8.52}$$

$$n_k = \frac{1}{e^{\beta\hbar\omega_k} - 1}. \tag{8.53}$$

Since the phonon occupation numbers n_k vanish in the limit of zero temperature, Δ_0 denotes the effective tunneling amplitude at $T = 0$; at finite temperature, thermal lattice vibrations cause a further reduction by the Debye-Waller-factor in (8.51).

The self-energy contributions $\Sigma_1, \Sigma_2, ...$ are expressed most conveniently in terms of the phase

$$\varphi(t) = \sum_k u_k^2 \left(\coth(\hbar\beta\omega_k/2)\cos(\omega_k t) - i\sin(\omega_k t)\right), \tag{8.54}$$

which may be rewritten as

$$\varphi(t) = \sum_k u_k^2 \frac{\cos(\omega_k(i\hbar\beta/2 + t))}{\sinh(\beta\hbar\omega/2)}. \tag{8.55}$$

Finally we note the Laplace transform

$$\varphi(z) = -\sum_k u_k^2 \left[\frac{n_k}{z + \omega_k} + \frac{1 + n_k}{z - \omega_k}\right] \tag{8.56}$$

and the spectral function

$$\varphi''(\omega) = \pi \sum_k u_k^2 \left[n_k \delta(\omega + \omega_k) + (1 + n_k)\delta(\omega - \omega_k)\right]. \tag{8.57}$$

When we insert (E6) and the phase φ, we find for (8.35)

$$\Sigma_1(t - t') = \frac{\tilde{\Delta}_0^2}{\hbar^2}\left(e^{\varphi(t-t')} - 1\right). \tag{8.58}$$

The second cumulant is determined by (E6,E7) and (E11,E12), and reads

$$\langle \delta K(t, \tau) \delta K(\tau', t') \rangle = (\tilde{\Delta}_0^4/\hbar^4) e^{\varphi(t-\tau) + \varphi(\tau'-t')} (\cosh(\Phi) - 1). \tag{8.59}$$

After removing the reducible part we have

$$\Sigma_2(t - t') = -\frac{\tilde{\Delta}_0^4}{\hbar^4} \int_{t'}^{t} d\tau \int_{t'}^{\tau} d\tau' G_1(\tau - \tau')$$
$$\times\ e^{\varphi(t-\tau) + \varphi(\tau'-t')}[\cosh(\Phi) - 1]. \tag{8.60}$$

Equations (8.50–8.60) are valid for any bath spectrum which vanishes sufficiently fast at zero frequency. To proceed further, we specify the summation over the heat bath by inserting the spectral density (8.7).

When we replace the sum over phonon modes in the usual way by an integral,

$$\sum_k u_k^2 \ldots \rightarrow 2\tilde{\alpha}(\hbar^2/k_B^2) \int_0^{\omega_D} d\omega\, \omega\ldots, \tag{8.61}$$

we obtain for the phase

$$\varphi(t) = (\hbar^2/k_B^2)\tilde{\alpha} \int_0^{\omega_D} d\omega\, \frac{2\omega}{\sinh(\beta\hbar\omega/2)} \cos(\omega(t + i\hbar\beta/2)). \tag{8.62}$$

Owing to the vanishing spectral density at zero frequency, the phase goes to zero in the long-time limit, $\varphi(t) \to 0$ for $t \to \infty$. Accordingly, the time-dependent parts of the self-energy tend towards zero as well,

$$\lim_{t\to\infty} \Sigma_1(t) = 0 = \lim_{t\to\infty} \Sigma_2(t), \tag{8.63}$$

whereas the first term Σ_0 is constant at all times.

Markov Approximation

In order to find the poles of (8.29), we have to solve $z + \Sigma(z) = 0$. The self-energy terms $\Sigma_1 + \Sigma_2 + \ldots$ are found to depend only weakly on frequency and thus may be evaluated at the bare poles $\pm\tilde{\Delta}_0$ or, even more simply, at zero frequency. In the latter case $\Sigma_1(z=0)$, $\Sigma_2(z=0)$, and so on, are purely imaginary and may be considered as damping rates.

8.5 The Damping Rate Γ_1

In this section, we derive the temperature dependence of the rate

$$\Gamma_1 = \Im \Sigma_1(z=0). \tag{8.64}$$

First, we expand the rate in terms of the coupling strength $\tilde{\alpha}$, which is particularly useful for the weak-damping or low-temperature limit. In the opposite case the rate may be evaluated by saddle-point integration. In Appendix E, we derive rigorous upper and lower bounds for Γ_1, which tend towards the lowest-order term of the perturbation series at low temperatures, and confirm the validity of the saddle-point integration at high temperatures.

Perturbation Series

According to (8.36), we have to evaluate the frequency dependent damping function

$$\Sigma_1(z) = i \int_0^\infty dt\, e^{izt} \frac{\tilde{\Delta}_0^2}{\hbar^2} \left(e^{\varphi(t)} - 1\right). \tag{8.65}$$

When we expand the exponential $e^{\varphi(t)}$ and take the Laplace transform of each term, we obtain

164   8. Small-Polaron Approach

$$\Sigma_1(z) = \frac{\tilde{\Delta}_0^2}{\hbar^2} \sum_{n=1}^{\infty} \frac{1}{n!} \kappa_n(z), \tag{8.66}$$

where $\kappa_n(z)$ is the transform of $\varphi(t)^n$,

$$\kappa_n(z) = \mathrm{i} \int_0^\infty \mathrm{d}t\, \mathrm{e}^{\mathrm{i}zt} \varphi(t)^n. \tag{8.67}$$

First we discuss the case of very low temperatures $k_\mathrm{B}T \leq \tilde{\Delta}_0$, where the series (8.66) may be restricted to its first term. At the bare poles $\pm\tilde{\Delta}_0/\hbar$, the self-energy $\Sigma_1(z)$ takes the value

$$\Sigma_1(\pm\tilde{\Delta}_0/\hbar) = \frac{\tilde{\Delta}_0^2}{\hbar^2} \varphi(\pm\tilde{\Delta}_0/\hbar). \tag{8.68}$$

When we insert (8.57) and the spectral density (8.7), we find for the damping rates of the poles at $\pm\tilde{\Delta}_0/\hbar$

$$\Im\Sigma_1(\mp\tilde{\Delta}_0/\hbar) = \pm(2\pi/\hbar)(\tilde{\alpha}/k_\mathrm{B}^2)\tilde{\Delta}_0^3(\mathrm{e}^{\pm\beta\tilde{\Delta}_0} - 1)^{-1}. \tag{8.69}$$

The fact that the two rates differ by a factor $\mathrm{e}^{\beta\tilde{\Delta}_0}$,

$$\Im\Sigma_1(\tilde{\Delta}_0/\hbar) = \mathrm{e}^{\beta\tilde{\Delta}_0}\Im\Sigma_1(-\tilde{\Delta}_0/\hbar), \tag{8.70}$$

merely shows that the upper state of the two-level system decays faster than the lower. Equation (8.70) yields the usual ratio of upward and downward scattering probabilities for a two-level system.

Destruction of phase coherence involves the average of the decay rates given in (8.69), yielding the well-known first-order rate

$$\varGamma_1 = \frac{\pi}{\hbar}(\tilde{\alpha}/k_\mathrm{B}^2)\tilde{\Delta}_0^3 \coth(\tilde{\Delta}_0/2k_\mathrm{B}T). \tag{8.71}$$

Now we turn to the high-temperature case $k_\mathrm{B}T \gg \tilde{\Delta}_0$, where the whole series (8.66) may contribute significantly to the rate. It proves convenient to rewrite the phase (8.62) as

$$\varphi(t) = 8\tilde{\alpha}T^2 \int_0^{\beta\hbar\omega_\mathrm{D}/2} \mathrm{d}x\, c(x) \cos[x(2tk_\mathrm{B}T/\hbar + \mathrm{i})], \tag{8.72}$$

where we have substituted the integration variable in (8.62) by $x = \beta\hbar\omega/2$ and defined

$$c(x) = \frac{x}{\sinh(x)}. \tag{8.73}$$

After we have inserted (8.72) and let $\Im z \to 0$, the time integration results in a delta function in the imaginary part of (8.67),

$$\Re \int_0^\infty \mathrm{d}t\, \mathrm{e}^{\mathrm{i}zt} \prod_{j=1}^n \cos[x_j(2t/\hbar\beta + \mathrm{i})]$$

$$= \frac{\pi\hbar\beta}{2} \sum_{\{\pm\}} 2^{-n} \delta\left(\beta\hbar z/2 \pm x_1 \pm \ldots \pm x_n\right) e^{\beta\hbar z/2}, \tag{8.74}$$

the sum running over all 2^n configurations of the signs.

The most interesting features of the correlation spectrum occur for frequencies of the order of $\tilde{\Delta}_0/\hbar$ or smaller. Since we assume $\tilde{\Delta}_0 \ll k_B T$, the quantity $\hbar z/2k_B T$ may be considered small compared with unity; it is clear from (8.74) that all $\kappa_n(z)$, and hence $\Sigma_1(\omega)$, depend only little on frequency. Accordingly, in the most relevant range $\beta\hbar z \ll 1$ we may evaluate (8.67) at zero frequency and take

$$\kappa_n(z \to 0) = \frac{\pi\hbar}{2k_B T} (4\tilde{\alpha}T^2)^n a_n, \tag{8.75}$$

where the coefficients a_n are given by

$$a_n = \lim_{\epsilon \to 0} \left(\prod_{j=1}^{n} \int_0^{\beta\hbar\omega_D/2} \mathrm{d}x_j\, c(x_j) \right) \sum_{\{\pm\}} \delta(\epsilon \pm x_1 \ldots \pm x_n). \tag{8.76}$$

(Note that the real part of $\Sigma_1(z=0)$ vanishes.)

At temperatures well below the Debye temperature $\Theta = \hbar\omega_D/k_B$, the upper limit of the integrals in (8.76) may be replaced by infinity. The first term then reads

$$a_1 = 1; \tag{8.77}$$

for $n \geq 2$ we perform the integration over x_n and obtain

$$a_n = \left(\prod_{j=1}^{n-1} \int_0^\infty \mathrm{d}x_j\, c(x_j) \right) \sum_{\{\pm\}} c(\pm x_1 \ldots \pm x_{n-1}), \tag{8.78}$$

where the sum runs over the remaining $(n-1)$ signs.

When we insert (8.75–8.78) in (8.66), we obtain

$$\Im\Sigma_1(\omega = 0) = \frac{\tilde{\Delta}_0^2}{\hbar^2} \sum_{n=1}^\infty \frac{1}{n!} \kappa_n(0) \equiv \Gamma_1. \tag{8.79}$$

Finally, we give explicitly the lowest-order terms of the damping rate,

$$\Gamma_1 = \frac{2\pi}{\hbar} \tilde{\Delta}_0^2 (\tilde{\alpha}/k_B) T \left(1 + \frac{2\pi^2}{3} \tilde{\alpha}T^2 + \ldots \right). \tag{8.80}$$

For $\tilde{\Delta}_0 \ll k_B T$, one has $\coth(\tilde{\Delta}_0/2k_B T) \approx 2k_B T/\tilde{\Delta}_0$, and finds the first-order term to be equal to (8.71). This term prevails at low enough temperatures satisfying $\tilde{\alpha}T^2 \ll 1$, whereas in the opposite limit higher-order contributions are dominant.

Starting from (8.75–8.79) rigorous upper and lower bounds are derived in Appendix E.

Saddle-Point Integration

Here we consider the overdamped or high-temperature limit $\tilde{\alpha}T^2 \gg 1$, corresponding to $\hbar\Gamma_1 \gg \tilde{\Delta}_0$. The imaginary part of the Fourier transform (8.65) at zero frequency may be replaced by an integral going from $-\infty$ to ∞,

$$\Gamma_1 = \frac{1}{2}\frac{\tilde{\Delta}_0^2}{\hbar^2}\int_{-\infty}^{\infty} dt \left(e^{\varphi(t)} - 1\right). \tag{8.81}$$

In the strong-coupling limit, this expression for the damping rate can be evaluated by distorting appropriately the contour of the time integral in the complex plane.

The phase $\varphi(t)$ is found to be stationary at $t_s = -i\hbar\beta/2$; a power series expansion in terms of $\tau = (t - t_s)$ yields

$$\varphi(\tau) = \varphi^{(0)} - \tfrac{1}{2}\varphi^{(2)}\tau^2 + \tfrac{1}{24}\varphi^{(4)}\tau^4 + \dots \tag{8.82}$$

with the moments

$$\varphi^{(n)} = \sum_k u_k^2 \frac{\omega_k^n}{\sinh(\hbar\omega_k/2k_B T)}. \tag{8.83}$$

After expanding the exponential $e^{\varphi(\tau)}$ in powers of the fourth- and higher order terms we are left with Gaussian integrals,

$$e^{\varphi^{(0)}}\int_{-\infty}^{\infty} d\tau e^{-\tfrac{1}{2}\varphi^{(2)}\tau^2}\left(1 + \frac{1}{4!}\varphi^{(4)}\tau^4 + \dots\right), \tag{8.84}$$

which result in the damping rate

$$\Gamma_1 = \frac{\tilde{\Delta}_0^2}{\hbar^2}\sqrt{\frac{\pi}{2\varphi^{(2)}}}e^{\varphi^{(0)}}\left(1 + \frac{1}{8}\frac{\varphi^{(4)}}{\varphi^{(2)2}} + \dots\right). \tag{8.85}$$

Here a word concerning the term -1 appearing in (8.81) is in order. The function $\Sigma_1(t) = \tilde{\Delta}_0^2(e^{\varphi(t)} - 1)$ has been defined such that its spectrum does not contain a delta peak at zero frequency. In the calculation of the Fourier transform $\Sigma_1''(\omega)$, both $e^{\varphi(t)}$ and "1" give rise to a contribution $\pi\delta(\omega)$, which cancel because of their opposite signs. Thus the spectrum $\Sigma_1''(\omega)$ is a smooth function of frequency, and may well be replaced by its value at $\omega = 0$, namely $\Gamma_1 = \Sigma_1''(\omega = 0)$.

When we truncate the series (8.84) at finite order, we obtain a self-energy spectrum that behaves regularly at zero frequency. Thus there is no singularity to be subtracted from the spectrum at zero frequency. The regular part, however, is given by (8.85). For a more detailed discussion see p. 177 and Mahan's book [163].

Temperature Dependence

The expression for the damping rate Γ_1, (8.85) still involves sums over phonon modes, which we are going to evaluate for the limiting cases $T \ll \Theta$ and $\Theta \ll T$; the Debye temperature $\Theta = \hbar\omega_D/k_B$ defines the cut-off of the frequency integrals. (We loosely follow Holstein [86] and Mahan [163].)

8.5 The Damping Rate Γ_1

$T \ll \Theta$. In this case we may replace the cut-off by infinity, as in the perturbative approach. When we perform the integration in the exponent of (8.51), we obtain with (8.61) for the tunnel energy

$$\tilde{\Delta}_0 = \Delta_0 e^{-\frac{1}{3}\pi^2 \tilde{\alpha} T^2}, \tag{8.86}$$

and for the phase

$$\varphi^{(n)} = 8\tilde{\alpha} T^2 (2k_B T/\hbar)^n I_{n+1}, \tag{8.87}$$

with the integral

$$I_n = \int_0^\infty dx \frac{x^n}{\sinh(x)}; \tag{8.88}$$

here we will need in particular [176]

$$I_1 = \tfrac{1}{4}\pi^2, \qquad I_3 = \tfrac{1}{8}\pi^4, \qquad I_5 = \tfrac{1}{4}\pi^6. \tag{8.89}$$

Thus we find for the first expansion coefficients of the time-dependent exponential

$$\varphi^{(0)} = 2\pi^2 \tilde{\alpha} T^2, \tag{8.90}$$

$$\varphi^{(2)} = 4\pi^4 (k_B/\hbar)^2 \tilde{\alpha} T^4, \tag{8.91}$$

$$\varphi^{(4)} = 32\pi^6 (k_B/\hbar)^4 \tilde{\alpha} T^6; \tag{8.92}$$

the rate with lowest-order corrections (8.85) reads

$$\Gamma_1 = \frac{\Delta_0^2}{\sqrt{8\pi^3 \tilde{\alpha} \hbar k_B T^2}} e^{\frac{4}{3}\pi^2 \tilde{\alpha} T^2} \left(1 + \frac{1}{16\pi^2 \tilde{\alpha} T^2} + \ldots\right) \qquad (T \ll \Theta). \tag{8.93}$$

From the correction terms in brackets, it is obvious that the saddle-point approximation breaks down in the weak-coupling, or low-temperature, range $\tilde{\alpha} T^2 < 1$. Now we turn to temperatures well beyond the Debye value Θ.

$\Theta \ll T$. With the above definitions, we rewrite the time-independent exponential of the saddle-point integration as

$$\tilde{\Delta}_0^2 e^{\varphi^{(0)}} = \Delta_b^2 \exp\left[-\sum_k u_k^2 \left(\coth(\beta \hbar \omega_k/2) - \sinh(\beta \hbar \omega_k/2)^{-1}\right)\right]. \tag{8.94}$$

For thermal phonons the quantities $x_k = \beta \hbar \omega_k/2$ are small compared with unity; hence we may replace the hyperbolic functions by the lowest-order terms of the series

$$\begin{aligned}\coth(x_k) &= 1/x_k + \tfrac{1}{2}x_k + O(x_k^3) \\ \sinh(x_k)^{-1} &= 1/x_k + O(x_k^2)\end{aligned} \tag{8.95}$$

and find

$$\tilde{\Delta}_0^2 e^{\varphi^{(0)}} = \Delta_b^2 e^{-V/k_B T}, \tag{8.96}$$

where we have defined

$$V = \tfrac{1}{2}\sum_k u_k^2 \hbar\omega_k. \tag{8.97}$$

By the same token we obtain with $\sinh(x_k)^{-1} \to 1/x_k$ in (8.83) the prefactor of the quadratic term in the exponential

$$\varphi^{(2)} = \frac{2k_B T}{\hbar}\sum_k u_k^2 \omega_k. \tag{8.98}$$

We replace the sum over phonon modes according to (8.61); thus we find in a straightforward fashion the activation energy

$$V = \tfrac{1}{3}k_B \tilde{\alpha}\Theta^3 \tag{8.99}$$

and the parameter of the Gaussian integral

$$\varphi^{(2)} = \tfrac{4}{3}\tilde{\alpha}(k_B/\hbar)^2 \Theta^3 T. \tag{8.100}$$

In the same way we obtain the fourth-order coefficient of the phase

$$\varphi^{(4)} = \tfrac{4}{5}\tilde{\alpha}(k_B/\hbar)^4 \Theta^5 T. \tag{8.101}$$

Now we insert (8.99) and (8.100) in the expression for the high-temperature rate, (8.85), and calculate the lowest-order correction term by means of (8.101),

$$\Gamma_1 = \frac{\Delta_b^2}{\hbar}\sqrt{\frac{3\pi}{8\tilde{\alpha}k_B^2\Theta^3 T}}\,e^{-V/T}\left(1 + \frac{45}{32}\frac{1}{\tilde{\alpha}\Theta T}\right) \qquad (T > \Theta). \tag{8.102}$$

Note that saddle-point approximation requires $1 \ll \tilde{\alpha}\Theta T$ holding. The rate (8.102) shows activated behavior for $\Theta < T < V$; at still higher temperatures it decreases according to $\Gamma_1 \propto T^{-1/2}$.

8.6 Cross-Over to Incoherent Motion

According to (8.36), the first-order approximation is given by

$$\Sigma(z) = \Sigma_0(z) + \Sigma_1(z). \tag{8.103}$$

Since (8.21) is constant in time, the first term exhibits a pole $\Sigma_0(z) = -(\tilde{\Delta}_0^2/\hbar^2)z^{-1}$ at $z = 0$, according to (8.30). We are mainly interested in the long-time or low-frequency behavior of the correlation function. The second term $\Sigma_1(t)$ decays sufficiently fast in the long-time limit; since $\Sigma_1(z)$ is a smooth function of frequency, it may be replaced by its value at $z = 0$. [See the discussion below (8.74).]

When we insert (8.30) in (8.36), we obtain the Laplace transform

$$G_1(z) = -\frac{z}{z^2 + iz\Gamma_1 - \tilde{\Delta}_0^2/\hbar^2}. \tag{8.104}$$

Note that with (8.17), the correlation spectrum vanishes at zero frequency, $G_1''(\omega) = 0$ for $\omega = 0$. The roots of the second-order polynomial in the denominator of (8.28) are easily calculated, yielding two poles in the lower part of the complex plane,

$$-i\tfrac{1}{2}\Gamma_1 \pm \sqrt{\tilde{\Delta}_0^2/\hbar^2 - \tfrac{1}{4}\Gamma_1^2}. \tag{8.105}$$

Both the pole structure and the time-dependent function $G(t)$ are quite different for the cases $\hbar\Gamma_1 < 2\tilde{\Delta}_0$ and $\hbar\Gamma_1 > 2\tilde{\Delta}_0$.

(a) Coherent tunneling: $\hbar\Gamma_1 < 2\tilde{\Delta}_0$. In this case (8.105) provides a pair of complex conjugate poles $-i\Gamma \pm E/\hbar$, with damping rate and resonance energy

$$\Gamma = \tfrac{1}{2}\Gamma_1 \qquad E = \sqrt{\tilde{\Delta}_0^2 - \hbar^2\Gamma^2}. \tag{8.106}$$

Laplace back-transformation of (8.104) yields for positive times

$$G_1(t) = e^{-\Gamma t}\cos(Et/\hbar - \delta)\cos(\delta)^{-1}, \tag{8.107}$$

with $\tan(\delta) = \Gamma/E$ varying from 0 to $\tfrac{\pi}{2}$. The two-time correlation function $G(t)$ exhibits damped oscillations with frequency $E/\hbar$. For later convenience, we note the correlation spectrum

$$G_1''(\omega) = \frac{\hbar\omega}{2E}\frac{\Gamma}{(\omega - E/\hbar)^2 + \Gamma^2} - \frac{\hbar\omega}{2E}\frac{\Gamma}{(\omega + E/\hbar)^2 + \Gamma^2}, \tag{8.108}$$

which shows the resonances at $\pm E/\hbar$ and vanishes at $\omega = 0$.

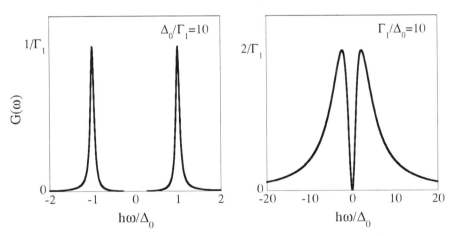

Fig. 8.1. Correlation spectrum $G_1(\omega)$ for $\Gamma_1/\tilde{\Delta}_0 = 1/10$ and $\Gamma_1/\tilde{\Delta}_0 = 10$, corresponding to the cases of weak and strong coupling, respectively. Both spectra vanish at zero frequency

(b) The aperiodic case $\hbar\Gamma_1 = 2\tilde{\Delta}_0$. At this particular value of the damping rate the resonance energy E vanishes; thus the two poles of (8.104) are identical and equal to $\frac{1}{2}$. By taking the back transform of (8.104) or by evaluating (8.107) in the limit $E \to 0$, we find

$$G_1(t) = e^{-\frac{1}{2}\Gamma_1 t}\left(1 - \tfrac{1}{2}\Gamma_1 t\right). \tag{8.109}$$

(c) Incoherent tunneling: $\hbar\Gamma_1 > 2\tilde{\Delta}_0$. Since the energy E is imaginary, it is convenient to define the real poles (8.105) as

$$\Gamma_\pm = \tfrac{1}{2}\Gamma_1 \pm \sqrt{\tfrac{1}{4}\Gamma_1^2 - \tilde{\Delta}_0^2/\hbar^2}, \tag{8.110}$$

which yields the time correlation function

$$G_1(t) = \frac{\Gamma_+}{\Gamma_+ - \Gamma_-}e^{-\Gamma_+ t} - \frac{\Gamma_-}{\Gamma_+ - \Gamma_-}e^{-\Gamma_- t}. \tag{8.111}$$

With (8.111) the spectral function may be rewritten as

$$G_1''(\omega) = \frac{\Gamma_+}{\Gamma_+ - \Gamma_-}\frac{\Gamma_+}{\omega^2 + \Gamma_+^2} - \frac{\Gamma_-}{\Gamma_+ - \Gamma_-}\frac{\Gamma_-}{\omega^2 + \Gamma_-^2}; \tag{8.112}$$

in accordance with (8.105), it vanishes at zero frequency.

In Fig. 8.1 we plot the spectrum $G_1''(\omega) = \Im G_1(\omega + i0)$ for two cases. For weak damping, $\hbar\Gamma_1 \ll 2\tilde{\Delta}_0$, the spectrum consists of two inelastic peaks at $\pm E/\hbar$; in the intermediate range it is small and vanishes at zero frequency. In the overdamped case, however, $G_1''(\omega)$ is dominated by a single Lorentzian with width $\Gamma_+ \approx \Gamma_1$, i.e., the first term in (8.112). Both amplitude and width of the second term are small; it causes the dip at zero frequency, thus satisfying the condition $G_1''(\omega = 0) = 0$.

For $\hbar\Gamma_1 < 2\tilde{\Delta}_0$, the correlation function $G_1(t)$ changes sign twice during a time interval $\hbar/E$; when approaching the aperiodic case $E \to 0$, the period $\hbar/E$ diverges. Both for $\hbar\Gamma_1 = 2\tilde{\Delta}_0$ and the overdamped case, $G_1(t)$ is positive at short times, then changes sign, and finally tends exponentially towards zero from below. [The small negative contribution to (8.111), decaying with rate Γ_-, prevails in the long-time limit.]

8.7 The Correction Term Σ_2

The dissipative two-state dynamics derived in the previous sections relies on the first-order expression Σ_1. Here we calculate next-order corrections by taking into account the self-energy contribution Σ_2. In this section we are going to evaluate the spectrum of (8.60) perturbatively, providing an appropriate low-temperature approximation.

Perturbation Theory: Low-Temperature Limit

In principle, one could write down a perturbation series for $\Sigma_2(z)$, similar to (8.66). Yet the six factors φ, which appear in the exponents in (8.60), and the first-order correlation function G_1 give rise to various convolutions in every term. Hence we are led to restrict the perturbation series for Σ_2 to the first finite contribution, which proves to be of second order in the small parameter $\tilde{\alpha}T^2$.

Since the phase φ is linear in the parameter $\tilde{\alpha}$, we expand $\Sigma_2(t)$ in terms of the former, and truncate at second order. The quantity $[\cosh(\Phi) - 1]$ in (8.60) is already quadratic in φ; thus we have to replace the remaining factors by their lowest-order approximation, i.e., e^φ by 1 and G_1 by G_0. Hence we obtain the lowest-order contribution

$$\Sigma_2(t-t') = -\frac{\tilde{\Delta}_0^4}{\hbar^4}\int_{t'}^{t}d\tau\int_{t'}^{\tau}d\tau'\frac{1}{2}\Phi(t,\tau,\tau',t')^2 G_0(\tau-\tau'). \tag{8.113}$$

We proceed as for Σ_1 in the previous section and concentrate on the low-frequency behavior of the Laplace transform $\Sigma_2(z)$.

After the definition of Φ is inserted in (8.113), the integrand consists of sixteen terms, which are shown schematically in Fig. 8.2. Up to a factor $(\tilde{\Delta}_0^4/\hbar^4)$, the first two terms read

$$\psi_a(t-t') = -\frac{1}{2}\varphi(t-t')^2\int_{t'}^{t}d\tau\int_{t'}^{\tau}d\tau' G_0(\tau-\tau'), \tag{8.114}$$

$$\psi_b(t-t') = -\frac{1}{2}\int_{t'}^{t}d\tau\int_{t'}^{\tau}d\tau'\varphi(\tau-\tau')^2 G_0(\tau-\tau'). \tag{8.115}$$

For convenience, we note the Laplace transform (8.16) for the integral of a function f

$$h(t) = \int_0^t d\tau f(\tau) \;:\; h(z) = \frac{i}{z}f(z) \tag{8.116}$$

and for the product of f and g

$$h(t) = f(t)g(t) \;:\; h(z) = \frac{i}{2\pi}\int dz' f(z')g(z-z'). \tag{8.117}$$

(With f and g analytical on the upper half of the complex plane, the integral over z' can be closed to a contour including only the poles of f.)

When we insert the expression for $\varphi(z)$ and perform the convolution integrals, we find for the Laplace transform of (8.115)

$$\psi_b(z) = \frac{1}{2z^2}\sum_{k,k'} u_k^2 u_{k'}^2\Big[(1+n_k)(1+n_{k'})G_0(z+\omega_k+\omega_{k'})$$
$$+2n_k(1+n_{k'})G_0(z-\omega_k+\omega_{k'}) + n_k n_{k'} G_0(z-\omega_k-\omega_{k'})\Big]; \tag{8.118}$$

for $\psi_a(z)$ we obtain a similar expression, but without the factor z^{-2}, and with the function $G_0(z)$ replaced everywhere by $z^{-2}G_0(z)$.

For simplicity we consider only the case $\tilde{\Delta}_0 \ll k_B T$. Since the most important contributions to the phonon sums stem from thermal frequencies $\omega_k \approx k_B T/\hbar$, we may discard the tunnel energy Δ_0 in the argument of G_0 and put $G_0(z) \approx -1/z$. When we evaluate the term in brackets at $z = 0$, we find that only the second term on the right-hand side of (8.118) gives a finite contribution. (Taking $z = 0$ instead of $z = \tilde{\Delta}_0/\hbar$ amounts to neglecting small terms of the order $(\tilde{\Delta}_0/k_B T)$.) In this way we obtain with

$$\Psi = \sum_{k,k'} u_k^2 u_{k'}^2 n_k (1 + n_{k'}) \delta(\omega_k - \omega_{k'}) \tag{8.119}$$

for the imaginary part of (8.118)

$$\Im \psi_b(z) = \pi z^{-2} \Psi. \tag{8.120}$$

Now we replace the sum over phonon modes according to (8.61), use $n_k(1 + n_k) = \frac{1}{4}\sinh(\beta\hbar\omega_k/2)^{-2}$, and substitute $x = \beta\hbar\omega_k/2$. With the integral (E15) we finally obtain

$$\Psi = \tfrac{4}{3}\pi^2 \tilde{\alpha}^2 (\hbar/k_B) T^3. \tag{8.121}$$

Now we turn to ψ_a. The double time integral in (8.114) reduces to the simpler expression

$$\psi_a(t - t') = -\frac{\hbar^2}{2\tilde{\Delta}_0^2} \varphi(t - t')^2 [1 - G_0(t - t')]. \tag{8.122}$$

With the Laplace transform of the term in brackets,

$$f_0(z) = z/(z^2 - \tilde{\Delta}_0^2/\hbar^2) - 1/z, \tag{8.123}$$

we find in analogy to (8.118)

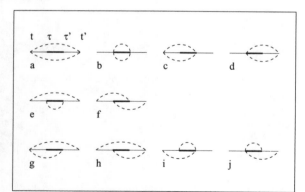

Fig. 8.2. Second-order diagrams arising from $\Sigma_2(t - t')$. The heavy line represents the spin propagator G_0, the dashed lines the phonon propagator φ

$$\psi_{\mathrm{a}}(z) = \frac{\hbar^2}{2\tilde{\Delta}_0^2} \sum_{k,k'} u_k^2 u_{k'}^2 \Big[(1+n_k)(1+n_{k'})f_0(z+\omega_k+\omega_{k'})$$

$$+2n_k(1+n_{k'})f_0(z-\omega_k+\omega_{k'}) + n_k n_{k'} f_0(z-\omega_k-\omega_{k'})\Big]. \quad (8.124)$$

As above, we replace $G_0(z)$ with $-1/z$, and find $f_0(z) = 0$ and thus $\psi_{\mathrm{a}}(z) = 0$. When we evaluate ψ_{a} at $z=0$ or $z = \tilde{\Delta}_0/\hbar$ and expand (8.124) in terms of $\tilde{\Delta}_0/k_{\mathrm{B}}T$, we obtain the leading term to be of the order $(\tilde{\Delta}_0/k_{\mathrm{B}}T)^2$, which is negligible.

The remaining terms of Σ_2 are insignificant compared with $\Im\psi_{\mathrm{b}}$. This may be understood by considering the propagators from t' to τ' and from τ to t in Fig. 8.2. Propagators without phonon lines yield a factor $1/z$, as occurs twice in (8.120). The function G_0 acts as a delta function; thus propagators with two phonon lines yield a factor $(\hbar/\tilde{\Delta}_0)$. Diagrams c and d of Fig. 8.2 comprise terms similar to $\Im\psi_{\mathrm{b}}$ and $\Im\psi_{\mathrm{a}}$ which cancel, however, because of opposite signs. All other terms involve additional factors ω_k^{-1}; when performing the phonon sums, they result in small factors $(\tilde{\Delta}_0/k_{\mathrm{B}}T)$.

When we insert the above results in Σ_2 and define

$$\gamma = \tfrac{4}{3}\pi^3 (1/k_{\mathrm{B}}\hbar)\tilde{\Delta}_0^2 \tilde{\alpha}^2 T^3, \quad (8.125)$$

we get the self-energy contribution

$$\Sigma_2(z) = \mathrm{i}\frac{\tilde{\Delta}_0^2}{\hbar^2 z^2}\gamma. \quad (8.126)$$

Comparison with (8.75) and (8.80) reveals that the constant γ is identical to the second-order term of the rate Γ_1.

The Correlation Spectrum $G_2''(\omega)$

The correction term Σ_2 modifies the correlation spectrum in several respects, as compared to (8.108). The constant in (8.126) causes a slight shift of the resonances, whereas the the second term gives rise to an additional feature of the spectral function $G_2''(\omega)$.

The correlation spectrum is obtained by inserting (8.126) in (8.29), and taking the imaginary part,

$$G_2''(\omega) = \frac{\omega^2(\omega^2\Gamma_1 + (\tilde{\Delta}_0^2/\hbar^2)\gamma)}{\left(\omega^2\Gamma_1 + (\tilde{\Delta}_0^2/\hbar^2)\gamma\right)^2 + \omega^2\left(\omega^2 - \tilde{\Delta}_0^2/\hbar^2\right)^2}. \quad (8.127)$$

When considering ω as a complex variable, the spectrum has six poles, three on the upper half and three on the lower half of the complex plane. In order to find approximate expressions for the poles, we rewrite the denominator as

$$(\omega^2 + \gamma^2)\Big[((\omega - E/\hbar)^2 + \Gamma^2)((\omega + E/\hbar)^2 + \Gamma^2) + \mathcal{R}_1\Big] + \mathcal{R}_2, \quad (8.128)$$

where we have defined the constants

8. Small-Polaron Approach

$$\mathcal{R}_1 = \gamma^2(\Gamma_1^2 - \gamma^2) \qquad \mathcal{R}_2 = 2\gamma^2(\Gamma_1\gamma + \gamma^2)\tilde{\Delta}_0^2/\hbar^2. \tag{8.129}$$

Resonance energy and damping rates are given by

$$\Gamma = \tfrac{1}{2}(\Gamma_1 + \gamma) \qquad E = \sqrt{\tilde{\Delta}_0^2 - \tfrac{1}{2}\hbar^2(\Gamma_1^2 - \gamma^2)}. \tag{8.130}$$

The validity of the perturbation analysis for the rates is restricted to the weak-damping regime where $\gamma \ll \Gamma_1 \ll \tilde{\Delta}_0/\hbar$. Both constants $\mathcal{R}_1$ and $\mathcal{R}_2$ are of fourth order in the rates and may therefore be neglected. Then (8.128) factorizes and provides the poles of the spectrum $G_2''(\omega)$.

The residues are most easily determined by writing the correlation spectrum as a sum of three different terms,

$$G_2''(\omega) = \frac{1}{2}\sum_{\pm}\left(\pm A\frac{\hbar\omega}{E} + B\right)\frac{\Gamma}{(\omega \mp E/\hbar)^2 + \Gamma^2} + C\frac{\gamma}{\omega^2 + \gamma^2}, \tag{8.131}$$

with residues A, B, and C. Those proportional to A and B have poles close to $\pm E$. The first one with residue A vanishes at zero frequency, in a similar way as in (8.108), whereas the second one with residue B is finite at $\omega = 0$. The third feature constitutes a quasi-elastic contribution with width γ and amplitude C. The sum rule $\int \mathrm{d}\omega G_2''(\omega) = \pi$ requires

$$A + B + C = 1. \tag{8.132}$$

When we rewrite (8.131) as a single fraction with the same denominator as (8.127) and set the corresponding numerators equal to each other, we obtain a set of equations for the residues A, B, C, whose solution reads to lowest order in $\hbar\gamma/\tilde{\Delta}_0$

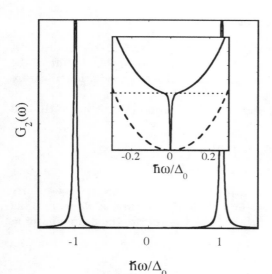

Fig. 8.3. Correlation spectra $G_1''(\omega)$ *(dashed line)* and $G_2''(\omega)$ *(full line)* as given by (8.108) and (8.131). The inset shows the behavior at small frequencies

$$A = 1 - \frac{\gamma}{\Gamma} + \frac{\hbar^2 \gamma^2}{\tilde{\Delta}_0^2}, \qquad B = \frac{\gamma}{\Gamma}, \qquad C = -\frac{\hbar^2 \gamma^2}{\tilde{\Delta}_0^2}. \tag{8.133}$$

Note that for low temperatures, Γ is given by the first term of the series (8.80); accordingly, we have $B = \gamma/\Gamma = \frac{1}{3}\pi^2 \tilde{\alpha} T^2$.

In Fig. 8.3 we plot both spectra $G_1''(\omega)$ and $G_2''(\omega)$ for $\hbar\Gamma/\tilde{\Delta}_0 = \frac{1}{20}$ and $\gamma/\Gamma = \frac{1}{5}$. In the weak-damping limit the first residues $A + B$ are close to unity and the spectrum is dominated by the main features at $\pm E/\hbar$. Close to these resonances, G_1 and G_2 are almost identical. Yet they differ significantly at small frequencies. After dropping small corrections, we find

$$G_2''(\omega) = \frac{\hbar^4 \omega^2 \Gamma}{\tilde{\Delta}_0^4} + \frac{\hbar^2 \gamma}{\tilde{\Delta}_0^2}\left(1 - \frac{\gamma^2}{\omega^2 + \gamma^2}\right) \qquad \text{for } |\omega| \ll E/\hbar. \tag{8.134}$$

The first term corresponds to G_1; it is negligible at small frequencies. In the inset of Fig. 8.3 we show the behavior of G_1 and G_2 close to $\omega = 0$. Clearly $G_2''(\omega)$ is determined by the second and third contributions in (8.134). The constant $\hbar^2\gamma/\tilde{\Delta}_0^2$ arises from the term proportional to B in (8.131). The negative quasi-elastic part with amplitude C is significant only at very small frequencies, $|\omega| \leq \gamma$, where it cancels the finite spectral density and assures $G_2''(\omega = 0) = 0$.

Hence the main effect of Σ_2 is a transfer of spectral density from high to low frequencies, resulting in an almost flat background between the two main resonances. As shown in the inset of Fig. 8.3, we have $G_2 > G_1$ at small frequency; for $\omega \geq E/\hbar$ we find $G_1 > G_2$.

Saddle-Point Integration: High-Temperature Limit

As in the case of the first-order contribution, saddle-point integration provides an appropriate means for evaluating the high-temperature rate. For $\tilde{\alpha} T^2 \gg 1$ both the oscillations of the phase φ and the factor $G_1(\tau - \tau')$ quickly destroy the coherence of the time evolution of (8.60). Hence the Laplace transform $\Sigma_2(z)$ is a smooth function of frequency and well approximated by its value at the undamped pole $z = \tilde{\Delta}_0$,

$$\Gamma_2 = \Im \Sigma_2(z = \tilde{\Delta}_0/\hbar). \tag{8.135}$$

To be more precise, there is always a singular contribution as in (8.126). Yet it proves to be of little consequence in the overdamped case, where the two poles arising from G_1 are purely imaginary. Retaining the $1/z^2$ contribution to (8.135) would result in a third relaxation pole with a rate comparable to Γ_1.

In order to evaluate the rate Γ_2, we approximate the six phase factors appearing in (8.60) according to

$$\varphi(\tau) = \varphi^{(0)} - \tfrac{1}{2}\varphi^{(2)}\tau^2; \tag{8.136}$$

inserting this in (E12) yields

$$\Phi(t,\tau,\tau',t') = \varphi^{(2)}(t-\tau)(\tau'-t'). \tag{8.137}$$

Owing to the exponentials $e^{\varphi(t-\tau)}$ and $e^{\varphi(\tau'-t')}$ in (8.60), the integrand is significant only when both time differences $(t-\tau)$ and $(\tau'-t')$ are shorter than $(\varphi^{(2)})^{-1/2}$. Hence Φ is smaller than unity, and the last factor of (8.60) may be expanded in terms of Φ,

$$(\cosh(\Phi)-1) = \tfrac{1}{2}\Phi^2 + \ldots, \tag{8.138}$$

and approximated by the quadratic term. Therefore

$$\Sigma_2(t-t') = \frac{1}{2}\frac{\tilde{\Delta}_0^4}{\hbar^4}\int_{t'}^{t}\mathrm{d}\tau\int_{t'}^{\tau}\mathrm{d}\tau' F(t-\tau)G_1(\tau-\tau')F(\tau'-t'), \tag{8.139}$$

with

$$F(\tau) = e^{\varphi^{(0)}}\varphi^{(2)}\tau^2 e^{-\tfrac{1}{2}\varphi^{(2)}\tau^2} \tag{8.140}$$

arising from (8.136–8.138).

When we note that (8.139) is a double convolution integral, we can easily calculate the Laplace transform $\Sigma_2(z)$. With

$$F(z) = \mathrm{i}e^{\varphi^{(0)}}\sqrt{\frac{\pi}{8\varphi^{(2)}}}\left(1-\frac{z^2}{\varphi^{(2)}}\right)\exp\left(-\frac{z^2}{2\varphi^{(2)}}\right) \tag{8.141}$$

(8.138) reads in frequency space

$$\Sigma_2(z) = \frac{1}{2}\frac{\tilde{\Delta}_0^4}{\hbar^4}F(z)^2 G_1(z). \tag{8.142}$$

In view of (8.135), we have to evaluate both $G_1(z)$ and $F(z)$ at $z=\tilde{\Delta}_0$. The latter function varies little with frequency and thus may be replaced with its value at $z=0$; with (8.85) we find

$$(\tilde{\Delta}_0^2/\hbar^2)F(z=0) = \tfrac{1}{2}\mathrm{i}\Gamma_1. \tag{8.143}$$

As to $G_1(z)$ we have to be more careful because of the dip in the spectrum $G''(\omega)$ at zero frequency. Yet it is clear from (8.112) that $G''(\omega) \approx 1/\Gamma_1$ except at very small and very large frequencies; thus we are led to take $G_1(z) = \mathrm{i}/\Gamma_1$ and find the second-order rate

$$\Gamma_2 = -\tfrac{1}{8}\Gamma_1. \tag{8.144}$$

When we insert (8.135, 8.144) in (8.48), we obtain a correlation spectrum very similar to (8.112), but with a slightly modified rate $\Gamma_1 + \Gamma_2 = \tfrac{7}{8}\Gamma_1$.

At a first glance it may seem surprising that in the high-temperature limit we obtain a negative contribution to the rate, whereas the above perturbation analysis enhanced the damping. This may be traced back to the different approximations for the quantity e^φ appearing in Σ_2. In the perturbative approach we have retained only the constant part and replaced both $e^{\varphi(t-\tau)}$ and $e^{\varphi(\tau'-t')}$ by unity (see (8.113)). On the other hand, in the high-temperature limit the quantity e^φ is determined by its fluctuating part; thus we have discarded the constant term and replaced $e^{\varphi(\tau)}$ with $\exp(\varphi^{(0)} - \tfrac{1}{2}\varphi^{(2)}\tau^2)$.

8.8 Discussion

Approximations

Choice of the initial state. In order to get rid of the additional inhomogeneity $I(t)$ in (8.15), we have chosen the density operator (8.27) for the thermal average. Accordingly, the pseudo-spin system occupies with equal probability the upper and lower level, and the statistical operator factorizes with respect to system and bath coordinates. The first property could be easily modified by choosing $\tilde{\rho} = \rho_S \rho$, with ρ_S involving spin operators only. The factorization property, however, is essential for the subsequent calculation. When we consider that $\tilde{\Delta}_0 \ll k_B T$ and that the coupling strength λ_k present an infinitely weak perturbation of the phonon mode k, the choice of the state (8.27) seems well-justified. In the literature this approximation is sometimes avoided by assuming the particle to be initially localized in one of the wells, which of course is just another choice of a product state $\rho_S \rho$.

Saddle-point integration. The saddle-point approximation for the integral (8.81) has been derived in many places [86,88]. After expanding the phase of the integrand $e^{\varphi(\tau)}$ in a power series in terms of τ^2, we have calculated next-order corrections to the usual Gaussian approximation. The correction terms in (8.93) involve inverse powers of $\tilde{\alpha} T^2$ and thus would seem negligible in the high-temperature limit.

In addition, in Appendix E we have derived upper and lower bounds $\mathcal{U}$ and $\mathcal{O}$ for the rate (8.81),

$$\mathcal{U} < \Gamma_1 < \mathcal{O}. \tag{8.145}$$

In the low-temperature limit, both tend towards the one-phonon rate. In the opposite case $\tilde{\alpha} T^2 \gg 1$, the upper bound (E29) differs from the saddle-point result merely by an insignificant factor; both $\mathcal{O}$ and Γ_1 comprise the same exponential factor. The lower bound $\mathcal{U}$ exhibits a similar behavior but with a smaller prefactor in the exponent; it probably could be improved by looking for a better lower limit for (E20).

At high temperature, or $\tilde{\alpha} T^2 \gg 1$, the result obtained from saddle-point integration lies between the upper and lower bounds given in (E29, E30), thus confirming the validity of this approximation and, in particular, that of the exponential increase of the rate Γ_1. Moreover, our rigorous bounds contradict the claim of Leggett et al. [15] that the saddle-point approximation is basically incorrect because of the neglect of the singular contribution of the spectrum. (Compare the discussion on p. 166.)

Cumulant expansion. Regarding the higher-order terms of the series for the self-energy in (8.29), we have calculated explicitly the second one, Σ_2, and given the expression (8.49) for the third one. At low temperatures, an expansion in powers of $\tilde{\alpha} T^2$ reveals that only Γ_1 comprises a linear term, and thus dominates the physics for $\tilde{\alpha} T^2 \ll 1$.

178   8. Small-Polaron Approach

In the overdamped limit $\tilde{\alpha}T^2 \gg 1$, the higher-order contributions to the damping rate are given by G_1 times a numerical factor; for the term of second order the latter reads $-\frac{1}{8}$, according to (8.144). When we consider the alternate sign (odd terms are positive, even negative) and note that the numerical factor seems to decrease with rising order, we are led to conjecture that the imaginary part of the series $\Sigma_1 + \Sigma_2 + ...$ differs little from the first-order value Γ_1.

Yet since the higher-order terms are not entirely negligible, a more rigorous evaluation of the correlations of the fluctuation operators ξ would seem desirable. In terms of the language coined in the path-integral approach to the Ohmic damping model [15], (8.144) means that the *non-interacting blip approximation* is not well-justified for the coupled phonon spectral density $J(\omega) \propto \omega^3$. Clearly, a proper treatment of the spin-boson model with cubic bath spectrum would require us to evaluate the full series in the denominator of (8.29), if possible in a more rigorous way than we have treated the second-order term.

Here a word concerning the scheme for constructing the successive self-energy contributions is in order. As an alternative to (8.41, 8.49) one might attempt to re-sum irreducible terms involving ξ operators only; the first terms of such a series read

$$\widetilde{\Sigma}_1(t-t') = \langle K(t-t') \rangle - \tilde{\Delta}_0^2/\hbar^2$$

$$\widetilde{\Sigma}_2(t-t') = -\int_{t'}^{t} d\tau \int_{t'}^{\tau} d\tau' \langle \delta K(t,\tau) \delta K(\tau',t') \rangle$$

$$\widetilde{\Sigma}_3(t-t') = \int_{t'}^{t} d\tau \int_{t'}^{\tau} d\tau' \int_{t'}^{\tau'} d\tau'' \int_{t'}^{\tau''} d\tau'''$$
$$\times \langle \delta K(t,\tau) \delta K(\tau',\tau'') \delta K(\tau''',t') \rangle. \quad (8.146)$$

The first one is identical to Σ_1, and it yields the rate Γ_1 as derived above.

Yet it is found that the rates arising from the higher-order terms are significantly larger than Γ_1, by factors of $\Gamma_1(\varphi^{(2)})^{-1/2}$. Owing to the alternate sign, the infinite series is still expected to yield the proper rate; truncation at finite order, however, does not provide a valid approximation scheme.

The Correction Term Σ_2

The first-order approximation for the self-energy, Σ_1, corresponds to the *non-interacting blip approximation* of the path-integral approach [15]. In the previous section, we have investigated the correction term Σ_2, and we have found the qualitative picture arising from NIBA to be essentially correct.

As a main difference, the $1/z^2$-singularity present in Σ_2 results in additional features of the correlation spectrum; in particular there is a background contribution at small frequencies $|\omega| \ll \tilde{\Delta}_0/\hbar$ and low temperatures.

This somewhat modifies the nature of the cross-over to relaxation. The spectrum $G_1(\omega)$ exhibits two inelastic features which broaden and move towards smaller frequencies until the respective poles merge on the imaginary axis. In second-order approximation, the spectrum comprises an additional flat contribution whose spectral weight increases with rising temperatures.

The above analysis is by no means complete. Yet it shows that in the overdamped regime, higher-order terms of the cumulant expansion (8.29) are not entirely negligible; the second-order correction Σ_2 leads to non-trivial modifications of the results obtained from NIBA.

Temperature Dependence

Cross-over temperature. The damping rate varies strongly with temperature. A most significant change occurs at the temperature where the rate Γ_1 is equal to twice the tunneling frequency $\tilde{\Delta}_0/\hbar$. Since this requires the quantity $\tilde{\alpha}T^2$ to be larger than unity, we use (8.93); when we neglect small corrections, the condition $2\tilde{\Delta}_0 = \hbar\Gamma_1$ defines a cross-over temperature through

$$T^* = \sqrt{\frac{3}{5\pi^2\tilde{\alpha}}} \log\left(\frac{\sqrt{\tilde{\alpha}}T^{*2}}{2\Delta_0}\right). \tag{8.147}$$

(Note that Δ_0 does not depend on temperature.) The temperature T^* differs by a logarithmic factor from

$$T_0 = \pi^{-1}(2\tilde{\alpha})^{-1/2}, \tag{8.148}$$

which has been used by several authors [88,91]. When we rewrite (8.147) in terms of T_0,

$$\frac{T^*}{T_0} = \sqrt{\frac{6}{5}} \log\left(\sqrt{\frac{6}{5}} \frac{T^{*2}}{2\pi T_0 \Delta_0}\right), \tag{8.149}$$

we find that for sufficiently small tunnel energy Δ_0, the temperature T^* may be significantly larger than T_0.

In physical terms, T_0 indicates the onset of the exponential decrease of the tunnel energy $\tilde{\Delta}_0$, according to (8.86); the rate exhibits a cross-over from the power-law behavior to the exponential dependence. Yet incoherent motion occurs only at the somewhat higher temperature T^*.

Beyond Debye temperature. The Debye temperature $\Theta = \hbar\omega_D/k_B$ defines the cut-off ω_D of the spectral density (8.7). Whereas for $T \ll \Theta$ the upper bound of the frequency integration could be replaced by infinity, in the opposite case $T > \Theta$ the integrals are determined by the cut-off frequency ω_D. [See, e.g., (8.100,8.101).]

According to Sect. 8.5, the damping rate Γ_1 is linear in temperature for $T < T_0$; well above T_0 it increases exponentially until the temperature approaches the Debye value Θ where the rate shows activated behavior and finally decreases. Clearly these laws are only to be observed far from the respective limits T_0 and Θ.

The above discussion is based on the relation $T_0 < \Theta$, which most physical systems seem to satisfy. Yet our results are easily extended to the case $\Theta < T_0$.

Markov Approximation

We conclude the discussion of the rate with a proviso concerning the Markov approximation in Sect. 8.4. In the incoherent regime, the rate may be much larger than temperature, $\hbar \Gamma_1 \gg k_B T$. If we evaluated (8.74) at frequencies about Γ_1, this would lead to a significant temperature factor which has been omitted by taking $\omega = 0$. This technical point may be put in physical terms by noting that relaxation with rate Γ_1 is driven by thermal phonons whose frequencies are much smaller than this rate.

Debye-Waller Factor

In the introductory section we have pointed out how perturbation theory in terms of the coupling parameter $\tilde{\alpha}$ fails to account for the renormalization of the bare tunnel energy. The exponential dependence on $\tilde{\alpha}$ of the tunnel energy (8.50,8.51) and (8.86) proves the strong-coupling character of the present approach; these reduction factors are often referred to as Debye-Waller factor with respect to phonon scattering. After expanding (8.86) in terms of $\tilde{\alpha}$ and truncating at first order, one recovers the perturbation theory result (8.2).

Relaxation in Glasses

In order to compare the results of this chapter with experimental data, we would have to take into account a finite asymmetry energy Δ and to average over both Δ_0 and Δ according to the distribution of the tunneling model. This program proves to be significantly more complicated for the small-polaron theory than for the mode-coupling approach of Chap. 8. Thus we merely address a few issues qualitatively.

Incoherent tunneling. The present approach confirms the cross-over to incoherent motion. The respective cross-over temperatures of the small-polaron and the mode-coupling theory differ by a logarithmic factor; they coincide for thermal two-level systems.

Yet the dependence of T^* on the tunnel energy affects the relaxation behavior. In Chap. 8 we have supposed that all two-level systems become incoherent at the same temperature, roughly equal to T_0. According to (8.149), the onset of incoherent motion is rather a continuous process starting at T_0; but for any temperature $T^* > T_0$ there are tunneling systems fulfilling the cross-over condition (8.149). Thus little can be said about the relaxation dynamics above T_0 to be expected from the present approach.

High-temperature rate. From data on sound propagation in amorphous solids one finds for T_0 values of about 8 Kelvin (see, e.g., Sect. 7.5). The Debye temperature in disordered materials is not well-defined; in the literature one finds values ranging from about 30 to 80 K. Thus the above laws are not likely to be observed experimentally as such; they merely indicate the general behavior of the rate.

We have assumed the cubic spectral density (8.7) to be valid up to the cut-off frequency ω_D. Since the length scale of tunneling systems in glasses, d, significantly exceeds the atomic distance, the limit of large wave lengths, or small phonon momentum, $kd \ll 1$, ceases to be valid well below the Debye frequency. For $kd \geq 1$ the phonon scattering matrix element depends on frequency, resulting in a strong reduction of the spectral density at high frequencies.

Hence the effective Debye temperature Θ is not much larger than T_0; according to (8.99) the activation energy V is then of the same order of magnitude as the Debye energy $k_\mathrm{B}\Theta$. Thus the various laws derived in Sect. 8.5 are not well-separated.

Comparison with Previous Work

Temperature dependence. The dynamics of a dissipative two-state system with super-Ohmic bath spectral function has been treated by diagrammatic perturbation theory [88], functional integral methods [15,91], and mode-coupling theory [95,128].

(i) In a Green's function approach with propagators $\langle \sigma_\alpha(t)\sigma_\beta(t')\rangle$ and $\alpha, \beta = \pm, z$, Pirc and Gosar [88] (PC) derived the dissipative dynamics of an asymmetric two-state system. Although these authors have taken into account a finite asymmetry energy Δ, we discuss the case $\Delta = 0$, in order to permit a comparison with our results. At low temperatures Pirc and Gosar find a damping rate which comprises the odd-order terms of our (8.79) only, and at high temperatures they obtain a relaxation rate which is identical to the lowest-order term of (8.85).

Both PC's and the present work rely on a power series expansion with respect to the fluctuation operators $\xi_\pm$ (8.19). PC calculate the rate in second order in $\xi_\pm$, which permits a rigorous statement only on the term linear in $\tilde{\alpha}T^2$. Our approach, however, includes contributions of fourth order in $\xi_\pm$, and thus yields exact results up to second order in $\tilde{\alpha}T^2$. From this we conclude that the rate conprises both odd and even terms. In fact this results is confirmed by higher-order perturbation theory with respect to the spin-phonon coupling.

(ii) Leggett et al. [15] have stated that a dissipative tunneling system with bath spectrum (8.7) would never get overdamped, and that its motion would always be given by weakly damped oscillations. The essential argument relies on the inequality

$$(\pi/k_{\rm B}^2)\tilde{\alpha}\tilde{\Delta}_0^3 \coth(\tilde{\Delta}_0/2k_{\rm B}T) \equiv \hbar\gamma_s \ll \tilde{\Delta}_0. \tag{8.150}$$

(We use our notation.) When we consider the exponential factor of $\tilde{\Delta}_0$, we find that condition (8.150) is always fulfilled, without restriction on $\tilde{\alpha}$ and T. The essentially exact result (8.66,8.75) reveals the weak-damping condition to read $\tilde{\alpha}T^2 \ll 1$; in this range γ_s is a good approximation for the rate Γ_1. In the opposite case, however, this rate is much larger than its first-order contribution, rendering the condition (8.150) meaningless.

Equation (8.150) expresses the fact that the first-order term (8.147) is much smaller than the tunneling frequency; yet the cross-over to incoherent tunneling is driven by higher-order terms. Thus for the cubic spectral density $J(\omega)$, the cross-over to incoherent motion is intimately connected to the exponential increase of the rate Γ_1.

According to [15] the singularity of $\Sigma(\omega)$ at $\omega = 0$ would imply oscillations at all temperatures, and incoherent motion would require $\Sigma(\omega = 0)$ to be finite. When calculating $\Sigma(\omega = 0)$ from $(\tilde{\Delta}_0^2/\hbar^2)e^{\varphi(\tau)}$, Holstein discarded the singular term (corresponding to our Σ_0) as an irrelevant diagonal process [86]. With the pole approximation $\omega + \Sigma(\omega) = 0$, we find this term Σ_0 to result in an insignificant modification of Holstein's result.

(iii) Using an imaginary-time path integral method, Grabert has investigated a tunneling system whose bath spectrum contains both linear and cubic parts [91]. In the incoherent regime the rate reads

$$\gamma_0\, e^{-\tfrac{1}{3}T^2/T_0^2} M(K, K + \tfrac{1}{2}, T^2/T_0^2), \tag{8.151}$$

which exhibits a cross-over at the temperature T_0. (The Kondo parameter K is given by the coupling constant α defined on p. 123.) For $T < T_0$, both the Kummer function $M(K, K + \tfrac{1}{2}, T^2/T_0^2)$ and the exponential factor are close to unity, and the rate is determined by Ohmic damping $\gamma_0 \propto T^{2K-1}$; in the opposite limit $T > T_0$, the phonon bath results in a strongly increasing rate.

In order to permit a comparison with our results, we set the Kondo parameter K equal to zero; then the Kummer function M is unity for all temperatures and $\gamma_0 = \Delta_0^2/(4\pi\hbar T)$. Grabert's rate $\gamma = \gamma_0 \exp(-\tfrac{1}{3}T^2/T_0^2)$ then shows an exponential *decrease* with rising temperatures, in contradiction with the lower bound $\mathcal{U}$ given by our formula (E30).

The discrepancy can be traced back to the form of the phase $\varphi(t)$; Grabert's function $\Lambda_{\rm ph}(t)$ is linked to our (8.62) by

$$\Lambda_{\rm ph}(t) = \sum_k u_k^2 \coth(\beta\hbar\omega_k/2) - \varphi(t - i\hbar\beta/2).$$

The first term on the right-hand side is trivial and has been absorbed in our definition of the tunneling energy $\tilde{\Delta}_0$; the time argument of the second term is essential. Using the phase $\Lambda_{\rm ph}(t)$ corresponds to neglecting the imaginary part $i\hbar\beta/2$ of the time argument in (8.62). Yet in our approach this would not seem justified. Note, however, that Grabert's approach to the phonon coupling already assumes relaxational motion, i.e., a sufficiently large value for the Kondo parameter K; thus our discussion based on the limit $K \to 0$ might be of little account.

(iv) The self-consistent equations arising from a mode-coupling approximation have been investigated by several authors [24,89,95,128]. The analytic solution presented in Chap. 7 has been obtained by applying a Markov approximation and neglecting the reactive part of the memory function; as a main result it shows a cross-over to relaxational motion at a temperature $T_0 = \pi^{-1}(2\tilde{\alpha})^{-1/2}$. The relaxation rate was found to be maximum at T_0 and to decrease as T^{-2} above T_0.

Comparing the results of the mode-coupling approximation with the present approach reveals that the high-temperature mode-coupling rate $\gamma_{\rm MC} = \Delta_{\rm b}^2/(\sqrt{\tilde{\alpha}}T^2)$ is identical to the prefactor of the result of the saddle-point integration (8.93); yet it lacks the exponential factor which is most significant at high temperatures.

Clearly this is a consequence of the neglect of the memory function's reactive part. Regarding the tunnel splitting, we expect that re-summing the perturbation series (whose linear term is given by (8.2)) would yield the exponential factor present in (8.50) and (8.51). Thus one might consider the tunnel energy Δ_0 used in the previous chapters as an effective quantity which already incorporates the polaron effect. Yet this argument does not apply to the rate which the present approach reveals to increase with rising temperature. (In the literature this effect is sometimes referred to as phonon-assisted tunneling [87,91,92].)

8.9 Summary and Conclusion

The theory presented in this chapter describes the dynamics of a dissipative two-state system in the range from zero temperature up to values well beyond the Debye temperature Θ. The findings concern both technical aspects of the spin-boson model and features of sound propagation in glasses.

(a) At temperature T^* as given in (8.147), a cross-over occurs from weakly damped oscillations to incoherent relaxation. In general T^* is larger than the temperature T_0, which indicates the break-down of perturbation theory.

(b) At high temperatures, the relaxation rate γ increases exponentially with temperature, as found earlier by Holstein and others. The diffusive polaron motion treated there corresponds to relaxational two-state dynamics in the present problem.

(c) The perturbation series for the correction terms Σ_2, Σ_3, ... start at 2nd, 3rd, ... order in terms of the coupling constant $\tilde{\alpha}T^2$. Thus they are negligible at low temperatures. Yet in the opposite limit $\tilde{\alpha}T^2 \gg 1$ they yield corrections to the rate that are of the same order of magnitude as the first-order result Γ_1. [See, e.g., (8.144).]

Now we turn to the consequences for the sound propagation in glasses, as discussed in the previous chapter. The two major issues concern the existence of a cross-over and the temperature dependence of the high-temperature relaxation rate.

(d) The small-polaron approach confirms the cross-over to incoherent motion derived from mode-coupling approximation. This finding supports our ascribing the change in sound velocity and attenuation at about 8 K to incoherent tunneling.

(e) Yet there is a significant discrepancy regarding the incoherent relaxation rate, which may be traced back to our neglecting the real part of the memory function in the mode-coupling approach. Taking into account the polaron effect yields a rate which increases with rising temperatures, whereas the mode-coupling result shows a power law decrease.

Obviously it would be most interesting to incorporate this additional temperature dependence of the rate in the description of tunneling in glasses. Yet when attempting to do so, one comes upon several snags. *First*, generalizing the results of this chapter to asymmetric two-level systems proves not to be a simple matter since particular attention has to be given to thermal systems whose asymmetry energy may be close to temperature. *Second*, the range of validity for the Debye phonon model is quite restricted in glasses. The Debye frequency is low, resulting in a complicated temperature dependence according to the discussion in the previous section. Moreover, above 15 K 'local oscillators' seem to overtake acoustic phonon modes [142,144,145]. Yet the temperature dependence derived in this chapter relies heavily on the spectral density of the heat bath; thus any serious attempt to go well beyond 10 K would require us to specify what should be substituted for the cubic spectrum of Debye phonons.

A. Spectral Representation

For Hermitian operators A, the symmetrized correlation function
$$C_A(t-t') = \tfrac{1}{2}\langle A(t)A(t') + A(t')A(t)\rangle \equiv (A(t)|A) \tag{A1}$$
is real and an even function of the time difference $(t-t')$ [174,132]. Its spectrum is obtained by Fourier transformation,
$$C''_A(\omega) = \frac{1}{2}\int_{-\infty}^{\infty} dt\, e^{i\omega t} C_A(t), \tag{A2}$$
$$C_A(t) = \frac{1}{\pi}\int_{-\infty}^{\infty} d\omega\, e^{-i\omega t} C''_A(\omega). \tag{A3}$$

For continued fractions the Laplace transform is more advantageous; note the factor i in the definition used here,
$$C_A(z) = i\int_0^{\infty} dt\, e^{izt} C_A(t). \tag{A4}$$

When we insert (A3) and exchange time and frequency integrations, we obtain the spectral representation
$$C_A(z) = \frac{1}{\pi}\int_{-\infty}^{\infty} d\omega\, \frac{C''(\omega)}{\omega - z} \quad (\mathrm{Im}\, z > 0). \tag{A5}$$

In the limit of real z, the real and imaginary parts of $C_A(z)$ read as
$$\lim_{\eta\to 0} C_A(\omega + i\eta) = C'_A(\omega) + iC''_A(\omega); \tag{A6}$$
the latter gives the correlation spectrum $C''_A(\omega)$, and is connected to the real part through a Kramers-Kronig relation
$$C'_A(\omega) = \frac{1}{\pi}\int_{-\infty}^{\infty} d\omega'\, \frac{C''_A(\omega')}{\omega' - \omega}. \tag{A7}$$

The mode-coupling approximation amounts to decoupling a correlation function of a composite operator
$$M(t) = (A(t)B(t)|AB) \tag{A8}$$
in products of time-dependent correlations

$$M_{\text{MC}}(t) = \frac{1}{2}\left(\langle A(t)A\rangle\langle B(t)B\rangle + \langle AA(t)\rangle\langle BB(t)\rangle\right). \tag{A9}$$

(This simple proceeding makes sense for $[A, B] = 0$ only.) In order to derive a closed set of equations, one needs to express M_{MC} through symmetrized functions C_A and C_B. Using the spectral representation of the Hamilton operator permits us to derive relations between the various spectra

$$\frac{1}{2}\int_{-\infty}^{\infty} dt\, e^{i\omega t}\langle A(t)A\rangle = \frac{2}{1+e^{-\beta\hbar\omega}}C_A''(\omega) \tag{A10}$$

$$\frac{1}{2}\int_{-\infty}^{\infty} dt\, e^{i\omega t}\langle AA(t)\rangle = \frac{2}{1+e^{\beta\hbar\omega}}C_A''(\omega), \tag{A11}$$

and to write the spectrum

$$M_{\text{MC}}''(\omega) = (C_A''(\omega) * C_B''(\omega)) \tag{A12}$$

as a weighted convolution integral

$$(C_A''(\omega) * C_B''(\omega)) \equiv \frac{1}{\pi}\int_{-\infty}^{\infty} d\omega'\, C_A''(\omega')C_B''(\omega-\omega')\frac{c(\omega)}{c(\omega')c(\omega-\omega')}. \tag{A13}$$

The temperature factors appearing in (A10,A11) are accounted for by

$$c(\omega) \equiv \cosh(\beta\hbar\omega/2). \tag{A14}$$

Clearly (A10–A13) are valid in thermal equilibrium only.

In two-state approximation, the dipole moment depends on the reduced position operator σ_z through $\boldsymbol{p} = \frac{1}{2}q\boldsymbol{d}\sigma_z$. Most theoretical work involves the symmetrized correlation function

$$G(t-t') = \frac{1}{2}\langle\sigma_z(t)\sigma_z(t') + \sigma_z(t')\sigma_z(t)\rangle, \tag{A15}$$

whereas experiments are concerned instead with the dielectric linear-response function with respect to the dipole moment $\boldsymbol{p}$. After averaging over the orientation of the dipole moment with respect to the external field, the response function reads

$$\chi(t-t') = \frac{1}{3V}\frac{p^2}{\hbar\epsilon_0}\sum_i i\langle[\sigma_z^i(t),\sigma_z^i(t')]\rangle; \tag{A16}$$

The spectra of correlation and response functions are linked by Eqs. (A10,A11) and fulfil a fluctuation-dissipation theorem

$$\chi''(\omega) = \frac{2np^2}{3\hbar\epsilon_0}\overline{G''(\omega)}\tanh(\beta\hbar\omega/2). \tag{A17}$$

Here have used the definition of the defect density $n = N/V = (1/V)\sum_i$ and the averaged correlation spectrum

$$\overline{G''(\omega)} = \frac{1}{N}\sum_i G_i''(\omega), \tag{A18}$$

the sum running over all N impurities.

B. Mori's Reduction Method

Time evolution of operators that do not obey simple commutation relations is an intricate matter in general. A projection method developed by Mori and Zwanzig permits an exact representation in terms of a few relevant operators [119,120]; the influence of the remaining degrees of freedom is accounted for by retarded memory functions.

Quantum mechanical operators of a many-body problem span a linear space of infinite dimension; by means of the Liouville operator $\mathcal{L}$, time evolution is determined by the linear equation

$$\hbar \partial_t A = \mathrm{i}[H, A] \equiv \mathrm{i}\hbar \mathcal{L} A, \tag{B1}$$

whose formal integral reads

$$A(t) = \mathrm{e}^{\mathrm{i}\mathcal{L}(t-t')} A(t'). \tag{B2}$$

When we define an appropriate inner product, this permits us to map the equation of motion (B1) onto a integro-differential equation on a low-dimensional subspace. The method relies on a proper choice of relevant operators and on the approximation for the memory function. After introducing the inner product, we discuss two schemes for applying the projection method.

B.1 Scalar Product

There are various functions satisfying the conditions for an inner product. The form proposed by Mori [119],

$$(A|B)_\mathrm{M} \equiv \frac{1}{\beta} \int_0^\beta \mathrm{d}\beta' \langle \mathrm{e}^{\beta' H} A \mathrm{e}^{-\beta' H} B \rangle, \tag{B3}$$

has been used extensively; angular brackets indicate the thermal average

$$\langle \ldots \rangle \equiv \mathrm{tr}(\mathrm{e}^{-\beta H} \ldots)/\mathrm{tr}(\mathrm{e}^{-\beta H}), \tag{B4}$$

and we restrict ourselves to Hermitian operators,

$$A = A^\dagger, \tag{B5}$$

the generalization being obvious. For time-dependent operators the Mori product is equal to Kubo's relaxation function [173,174]

$$\Phi_{AB}(t-t') = (A(t)|B(t'))_M; \tag{B6}$$

for classical variables it reduces to the simple correlation function $\langle A(t)B(t')\rangle$.

The algebra of Pauli matrices suggests another choice, namely the symmetrized correlation function

$$(A|B) \equiv \frac{1}{2}\langle AB + BA\rangle, \tag{B7}$$

which yields to a particularly simple static correlation matrix

$$(\sigma_\alpha|\sigma_\beta) = \delta_{\alpha\beta}. \tag{B8}$$

B.2 Correlation Matrix of Relevant Operators

First, we consider a single projection on a finite-dimensional space spanned by a set of relevant operators $\{A_\alpha\}$ [119]. In principle, they need neither be perpendicular to each other nor be related by the equation of motion. Yet often operators appearing in the Hamiltonian and their time derivatives prove to be important. We consider two-time correlation functions

$$C_{\alpha\beta}(t-t') \equiv (A_\alpha(t)|A_\beta(t')) \tag{B9}$$

as defined in (A1), and we assume the thermal expectation value to vanish

$$\langle A_\alpha \rangle = 0. \tag{B10}$$

The latter condition can always be achieved by subtracting the static part of each operator.

The scalar product (B7) defines both the static correlation matrix

$$\eta_{\alpha\beta} \equiv (A_\alpha|A_\beta) \tag{B11}$$

and the projections

$$\mathcal{P} \equiv \sum_{\alpha\beta} |A_\alpha)\eta_{\alpha\beta}^{-1}(A_\beta| \equiv 1 - \mathcal{Q}. \tag{B12}$$

In terms of the latter, the Liouville operator splits in four parts

$$\mathcal{L} = \mathcal{P}\mathcal{L}\mathcal{P} + \mathcal{P}\mathcal{L}\mathcal{Q} + \mathcal{Q}\mathcal{L}\mathcal{P} + \mathcal{Q}\mathcal{L}\mathcal{Q}. \tag{B13}$$

Insertion in the equation of motion

$$\partial_t C_{\alpha\beta}(t) = (i\mathcal{L}A_\alpha(t)|A_\beta) \tag{B14}$$

and formal integration of the inhomogeneous term yield the integro-differential equation

$$\partial_t C_{\alpha\beta}(t) = i\Omega_{\alpha\gamma}\eta_{\gamma\delta}^{-1}C_{\delta\beta}(t) - \int_0^t dt' M_{\alpha\gamma}(t-t')\eta_{\gamma\delta}^{-1}C_{\delta\beta}(t'), \tag{B15}$$

where the secular part

$$i\Omega_{\alpha\beta} \equiv (\dot{A}_\alpha|A_\beta) = \dot{C}_{\alpha\beta}(t=0) \tag{B16}$$

describes time evolution in the relevant space, and the memory matrix

$$M_{\alpha\beta}(t) \equiv (\mathcal{Q}\dot{A}_\alpha|e^{-i\mathcal{Q}\mathcal{L}\mathcal{Q}t/\hbar}\mathcal{Q}\dot{A}_\beta) \tag{B17}$$

accounts for that occurring the complement space. (We use the shorthand notation $\eta_{\alpha\beta}^{-1}$ for $(\eta^{-1})_{\alpha\beta}$.)

When the retarded part is given as a convolution, the equation of motion can be solved by Laplace transformation, resulting in the matrix equation

$$\mathbf{C}(z) = \eta \frac{-1}{z\eta + \Omega + \mathbf{M}(z)}\eta. \tag{B18}$$

B.3 Continued Fraction Expansion

Now we consider a different proceeding that is based on repeated application of one-dimensional projections, and that permits to construct an orthogonal set $\{A_i\}$ from a single operator A and its derivatives,

$$A, \dot{A}, \ddot{A}, \ldots \rightarrow A_0, A_1, A_2, \ldots \tag{B19}$$

$$(A_i|A_j) = \delta_{ij}(A_i|A_i). \tag{B20}$$

We start with a normalized operator A and the correlation function

$$C(t-t') = (A(t)|A(t')) \tag{B21}$$

with $C(t=0) = 1$. When we take $A_0 \equiv A$ and iterate

$$A_{j+1} \equiv \mathcal{Q}_j \dot{A}_j \tag{B22}$$

by means of the projections

$$\mathcal{P}_j \equiv \eta_j^{-1}|A_j)(A_j| \equiv 1 - \mathcal{Q}_j \tag{B23}$$

with the normalization constant

$$\eta_j \equiv (A_j|A_j), \tag{B24}$$

we obtain a set of operators A_j which are linearly independent of each other with respect to the scalar product (B7).

Time evolution of the corresponding normalized correlation function

$$C_j(t) \equiv \eta_j^{-1}(A_j(t)|A_j)_{(j)} \tag{B25}$$

is given by the projected Liouville operator

$$\mathcal{L}_{j+1} \equiv \mathcal{Q}_j \mathcal{L}_j = \prod_{k=0}^{j} \mathcal{Q}_k \mathcal{L}, \tag{B26}$$

which is indicated by a subscript j. Note that the operator A_j does not evolve into the subspace spanned by $A_0, A_1, \ldots, A_{j-1}$.

Upon Laplace transformation of (B21),

$$C(z) = (A|[\mathcal{L} - z]^{-1}|A) \tag{B27}$$

and repeated use of the resolvent identity

$$\mathcal{P}\frac{1}{z-\mathcal{L}}\mathcal{P} = \frac{\mathcal{P}}{z - \mathcal{P}\mathcal{L}\mathcal{P} + \mathcal{P}\mathcal{L}\mathcal{Q}\dfrac{1}{z - \mathcal{Q}\mathcal{L}\mathcal{Q}}\mathcal{Q}\mathcal{L}\mathcal{P}}, \tag{B28}$$

we obtain the recursion relation

$$C_j(z) = \frac{-1}{z + \omega_j + \Delta_{j+1}^2 C_{j+1}(z)}, \tag{B29}$$

where we have used

$$i\omega_j \equiv \dot{C}_j(t = 0) \tag{B30}$$

$$\Delta_{j+1}^2 \equiv \eta_{j+1}/\eta_j. \tag{B31}$$

Laplace back transformation of (B29) yields the integro-differential equation

$$\frac{\partial}{\partial t}C_j(t) = i\omega_j C_j(t) - \int_0^t dt' \Delta_{j+1}^2 C_{j+1}(t - t') C_j(t'), \tag{B32}$$

which exhibits how memory effects arise from higher-order correlation functions.

When $\mathcal{P}_j$ projects on a one-dimensional space, the frequencies (B30) vanishes for a Hermitian operator A,

$$\omega_j = 0. \tag{B33}$$

This is most obvious from (B30), since for $A = A^\dagger$ the correlation $C_j(t)$ is an even function of time with a slope that is necessarily zero at $t = 0$.

Finally we note a particular choice for the operators $\{A_\alpha\}$. If these are identical to the orthogonal set $A_0, ..., A_{j_0}$ constructed in Sect. B.2, the frequency matrix is finite just below and above the diagonal, $\Omega_{\alpha\beta} = 0$ for $|\alpha - \beta| \neq 1$, and the memory matrix consists of a single finite entry $M_{j_0 j_0}$.

B.4 Approximations

Both the correlation matrix (B18) and the continued fraction (B29) are formally exact. Yet they suggest different approximations; hence for a given problem, one or the other will prove the more appropriate.

As a general rule, one includes as many operators A_α in the correlation matrix that one can still calculate the frequency matrix Ω and the static correlations η. Accordingly, the continued fraction is pursued as long as the normalization constants η_j can be evaluated. When truncating at sufficiently low order, these static quantities can calculated exactly, or in well controlled approximations.

The complication arising from the many-body problem is then hidden in the memory functions; in order to derive explicit results, usually quite severe simplifications are required. Either the memory function is a smooth function of frequency and thus may be replaced with a constant, amounting to a Markov approximation, or it shows a more complicated frequency dependence, requiring a more detailed treatment. In the latter case it is evaluated in some decoupling scheme where time correlations of composite operators are approximated by products of correlations of single operators; as a most prominent example we note the mode-coupling approximation.

C. Average over Disorder

A basic assumption of the present description of defect crystals consists in the assumption of a random configuration of the impurities on the host lattice. This assumption determines the distribution for dipolar couplings $P(J_{ij})$, and for all quantities which are functions of these. Starting from $P(J_{ij})$, we derive the distribution functions for the quantities I_i and τ_i defined in (5.47) and (5.76).

C.1 Distribution of Dipolar Couplings J_{ij}

The dipolar interaction (3.5) depends both on the distance of two dipoles, R_{ij}, and on the relative orientation of the vectors $\boldsymbol{R}_{ij}, \boldsymbol{r}_i, \boldsymbol{r}_j$. (We use the notation defined in Chap. 4.) The angular dependence for a given distance R_{ij} is accounted for by allowing for both signs in the expression

$$\frac{1}{2}J_{ij} = \pm \frac{1}{4\pi\epsilon_0} \frac{p^2}{R_{ij}^3}. \tag{C1}$$

Consider a defect at site i. The remaining $N-1$ impurities are randomly arranged on the remaining $N_0 - 1$ sites; in the sequel, we neglect corrections of the order $1/N$ or $1/N_0$. When we assume a homogeneous impurity distribution, this fixes the distribution of the distance R_{ij}; with $\int \mathrm{d}R_{ij}^3 = V$ and

$$V^{-1}\mathrm{d}^3 R_{ij} = P(J_{ij})\mathrm{d}J_{ij}, \tag{C2}$$

we find the normalized distribution

$$P(J_{ij}) = \frac{1}{2}\frac{J_1 J_2}{J_2 - J_1}\frac{1}{J_{ij}^2} \qquad \text{for} \quad J_1 \leq |J_{ij}| \leq J_2. \tag{C3}$$

The defect density is given by $n = N/V$ and the density of lattice sites by $n_0 = N_0/V$.

The upper bound $\frac{1}{2}J_2 = (1/4\pi\epsilon_0)(p^2/R_0^3)$ is fixed by a minimum distance R_0 which we relate to the site density by taking

$$R_0 = \left(\frac{3}{4\pi}\frac{1}{n_0}\right)^{1/3}; \tag{C4}$$

we thus find
$$J_2 = \frac{2}{3\epsilon\epsilon_0} n_0 p^2. \tag{C5}$$

The lower bound J_1 depends on the inverse sample volume V. (Accordingly, the limit of infinite volume leads to a divergence of (C3) for $J \to J_1$; the larger the volume, the smaller the minimum coupling J_1, and the larger the probability of such weakly interacting pairs.) From the normalization condition $\int dJ P(J) = 1$ we get
$$J_1 = \frac{1}{N_0} J_2. \tag{C6}$$

The derivation of (C3–C6) is based on a continuous distribution for R_{ij}. For $R_0 \ll R_{ij}$ this assumption is well justified, whereas it breaks down for distances of the order of the lattice spacing a; in that case, one has to consider the actual spacings on a lattice, R_{ij}, resulting in discrete values for J_{ij}. (Compare Chap. 3.) The definition of R_0 is to some extent arbitrary; a different choice would result in a numerical factor of the order of unity.

For potassium chloride as the host crystal we find $R_0 = 2.46$ Å, which is smaller than the length of the unit cell, $a = 6.23$ Å, or the distance of nearest-neighbor potassium sites, 4.41 Å. Accordingly, the upper cut-off for the dipolar interaction, J_2, is almost six times larger than the actual maximum coupling of impurities on nearest-neighbor sites, J_{NN}. (Yet this discrepancy is of little relevance, since all quantities derived in Chap. 5 depend only on the average interaction cJ_2 to be discussed below.) When we insert the dipole moment of a Li impurity and the dielectric constant of KCl, we find $J_2/k_B = 1508$ K, and $J_{\mathrm{NN}} = 259$ K. The use of the macroscopic value for the dielectric ϵ is not well-justified on such short distances (cf. discussion on p. 28); in fact it is needed mainly for impurity pairs with average distance

$$R = \left(\frac{3}{4\pi}\frac{1}{n}\right)^{1/3}, \tag{C7}$$

which is by a factor $c^{-1/3}$ larger than R_0.

The average of a quantity F with respect to (C3) is given by
$$\overline{F(J_{ij})} \equiv \int dJ_{ij} P(J_{ij}) F(J_{ij}). \tag{C8}$$

Owing to the equal probability for both signs, the odd moments vanish
$$\overline{J_{ij}^{2n-1}} = 0 \quad \text{for } n \text{ integer}; \tag{C9}$$
the first even moments read
$$\overline{J_{ij}^2} = J_1 J_2, \tag{C10}$$
$$\overline{J_{ij}^4} = (J_1 J_2)^2 + \frac{1}{3} J_1 J_2 (J_2^2 - J_1^2). \tag{C11}$$

C.2 The Distribution Function $Q(I)$

The correlation functions derived in Chap. 5 depend on the sum I_i^2 rather than on the dipolar couplings individually; their distribution is quite different from, though not independent of, $P(J_{ij})$.

The average of F with respect to I,

$$\overline{F(I)} \equiv \int dI\, Q(I) F(I), \tag{C12}$$

is to be calculated by means of the distribution

$$Q(I) = \overline{\delta\left(I - \sqrt{\Sigma_j J_{ij}^2}\right)}, \tag{C13}$$

which does not seem to be easily connected to (C3). Evaluation of the moments

$$\overline{I^n} = \overline{(\sum_j J_{ij}^2)^{n/2}} \tag{C14}$$

is not simple either, except for the second one which reads with (C10)

$$\overline{I^2} = N J_1 J_2 = c J_2^2. \tag{C15}$$

Considering the fourth moment, we separate the diagonal contributions from the remainder,

$$\overline{I^4} = \sum_j \overline{J_{ij}^4} + \sum_{j,k \neq j} \overline{J_{ij}^2 J_{ik}^2}; \tag{C16}$$

the latter term is negligible, since moments with positive exponent are determined by the upper bound of the distribution (C3). Thus for small concentration $c \ll 1$ the even moments read

$$\overline{I^{2n}} = \frac{c}{2n-1} J_2^{2n} \quad \text{for } n > 0;. \tag{C17}$$

corrections involve powers of c and $1/N_0$. Neither the odd moments nor those with negative exponent $n < 0$ can be derived in the same way.

Consider the integral of (C13) from 0 to I,

$$\int_0^I dI'\, Q(I') = \overline{\Theta(I^2 - \Sigma_j J_{ij}^2)} \equiv Z(I), \tag{C18}$$

where Θ is the usual step function, and the average over J_{ij} involves all possible arrangements of the $N-1$ remaining impurities on the lattice. Equation (C18) is only the probability that the interaction energy (5.47) takes a value smaller than or equal to I; due to the strong dependence of the argument of the step function on distance, $J_{ij}^2 \propto R_{ij}^{-6}$, the sum over j is mainly determined by the nearest neighbor's contribution.

Using $J_{ij} = \pm J_2 (R_0/R_{ij})^3$, the condition imposed on the step function may be approximated as

$$R_{ij}^3 \geq R_0^3 \frac{J_2}{I} \qquad \text{for all } j. \tag{C19}$$

This condition, however, is identical to the requirement that none of the remaining $(N-1)$ defects be found in a volume $\frac{4\pi}{3} R_0^3 J_2/I$ about the site i. According to the binomial distribution, the chance of finding k impurities on n sites is

$$\binom{n}{k} c^k (1-c)^{n-k} \tag{C20}$$

where c is the probability for one site to be occupied. With (C19) there are $n = J_2/I - 1$ empty sites in the volume $\frac{4\pi}{3} R_{ij}^3$ with site i at its center; setting $k = 0$ we have

$$Z(I) = (1-c)^{J_2/I - 1}. \tag{C21}$$

Using $\log(1-c) \approx -c$ for $c \ll 1$, we finally obtain

$$Z(I) = e^{c(1 - J_2/I)}. \tag{C22}$$

Note that (C22) shows the correct behavior at the upper and lower bounds.
(i) $I = 0$ yields $Z = 0$: zero interaction energy I requires an infinite empty volume about site i; such a configuration occurs with zero probability.
(ii) The case $I = J_2$ gives $Z = 1$: According to (C19), the coupling energy of impurities on adjacent sites corresponds to a volume $\frac{4\pi}{3} R_0^3$ which is just the volume excluded by the impurity at site i; yet this volume is free of additional defects with probability $Z(J_2) = 1$.

Finally we take the derivative of $Z(I)$ and obtain the distribution function

$$Q(I) = \frac{cJ_2}{I^2} e^{c(1 - J_2/I)} \qquad c \ll 1, \tag{C23}$$

which is defined on the interval $0 \leq I \leq J_2$; by substituting $x \equiv I_c/I$, it is easy to show that (C23) is properly normalized,

$$\int \mathrm{d}I\, Q(I) = 1, \tag{C24}$$

and to derive, to lowest order in c, the inverse moments

$$\overline{I^{-n}} = \frac{n!}{c^n J_2^n} \qquad \text{for } n > 0. \tag{C25}$$

For evaluation of the remaining moments, we approximate (C23) by replacing the exponential with an appropriate step function

$$\tilde{Q}(I) = \frac{cJ_2}{I^2} \Theta(I - I_0) \qquad \text{for } c \ll 1, \tag{C26}$$

with the lower bound

$$I_0 \equiv \frac{c}{1+c} J_2 \tag{C27}$$

being fixed by the normalization condition $\int dI \tilde{Q}(I) = 1$. Now it is straightforward to integrate

$$\overline{I} = c \log\left(\frac{1+c}{c}\right) J_2, \tag{C28}$$

$$\overline{I^n} = \frac{c}{n-1} J_2^n \quad \text{for } n > 1. \tag{C29}$$

We stress that corrections to (C26) and to the moments (C28,C29) involve terms of the order $O(c^2)$; the even moments agree with the exact result (C15) and with (C17), which again is identical to the lowest order in c. For sufficiently dilute defect systems, we have $c \ll 1$ and thus may replace I_0 with

$$I_0 \to I_c = cJ_2. \tag{C30}$$

According to (C25,C28), the quantity I_c determines the most relevant moments of the dipolar interaction I. For this reason we refer to I_c as the average interaction energy; for convenience we note its expression in terms of the defect density n and dipole moment p,

$$I_c = \frac{2}{3\epsilon\epsilon_0} np^2. \tag{C31}$$

C.3 Distribution of τ_i

Besides the interaction energy I_i, the quantity

$$\tau_i^2 = \sum_{j \neq j_0} J_{ij}^2 \tag{C32}$$

is relevant for the parameter κ defined in (5.63); in analogy to (C13), its distribution function is given by

$$Q_1(\tau) = \overline{\delta\left(\tau - \sqrt{\Sigma_{j \neq j_0} J_{ij}^2}\right)}, \tag{C33}$$

where the restriction on the sum excludes the nearest neighbor j_0. As in (C18) we integrate from 0 to τ,

$$\int_0^\tau d\tau' Q_1(\tau') = \overline{\Theta(\tau^2 - \Sigma_{j \neq j_0} J_{ij}^2)} \equiv Z_1(\tau); \tag{C34}$$

the average covers all configurations of the remaining $N-1$ impurities that satisfy the condition of the step function. Owing to the strong dependence of the dipolar coupling on distance, $J_{ij}^2 \propto R_{ij}^{-6}$, the quantity τ_i is determined by the contribution of the next-nearest neighbor. (The nearest neighbor j_0 is excluded from the sum.) Thus the argument of the step function is positive as long as less than two impurities violate the condition (C19), i. e., if

C. Average over Disorder

$$R_{ij}^3 \geq R_0^3 \frac{J_2}{\tau} \quad \text{for all } j \text{ except } j_0. \tag{C35}$$

Equation (C35) is satisfied by any configuration where none or at the most one of the $(N-1)$ remaining defects is found in the volume $\frac{4\pi}{3} R_0^3 J_2/\tau$ about the site i. According to the binomial distribution, such configurations occur with a probability of

$$\binom{n}{0}(1-c)^n + \binom{n}{1}c(1-c)^{n-1} \tag{C36}$$

where the volume given offers $n \equiv (J_2/\tau - 1)$ empty lattice sites. Proceeding as above in (C22), we obtain

$$Z_1(\tau) = e^{c(1-J_2/\tau)} + c\left(\frac{J_2}{\tau} - 1\right)e^{c(2-J_2/\tau)}. \tag{C37}$$

The limit $\tau \to 0$ corresponds to an infinite volume occupied by a single impurity, which occurs with zero probability, $Z_1(0) = 0$. The opposite limit $Z_1 = 1$ is achieved by the upper bound for τ which is close to $J_2/2$.

The derivative with respect to τ finally gives the distribution function

$$Q_1(\tau) = \frac{I_c}{\tau^2}e^{c-I_c/\tau} + \left(\frac{I_c^2}{\tau^3} - (1+c)\frac{I_c}{\tau^2}\right)e^{2c-I_c/\tau}, \tag{C38}$$

with moments

$$\overline{\tau^n} \equiv \int d\tau\, \tau^n Q_1(\tau). \tag{C39}$$

When we substitute $x \equiv I_c/\tau$ and integrate from $x_{\min}$ to infinity, we find the first moments

$$\overline{\tau} = I_c \left(e^{2c-x_{\min}} + (e^c - (1+c)e^{2c})\text{Ei}(-x_{\min})\right) \tag{C40}$$

$$\overline{\tau^2} = I_c^2 \left((e^{2c} - e^c)\frac{e^{-x_{\min}}}{x_{\min}} + (e^c - 2e^{2c})\text{Ei}(-x_{\min})\right) \tag{C41}$$

At low concentrations $c \ll 1$, the condition $Z(I_c/x_{\min}) = 1$ yields the lower bound $x_{\min} \approx 2c$, and the exponential integral is given approximately by $\text{Ei}(-2c) \approx C + \log(2c)$, where $C \approx 0.577$ is Eulers constant; when we discard small terms of the order $c\log(c)$ in (C40), as well as $(C+\log 2)/\log c$ in (C41), we finally obtain

$$\overline{\tau} = I_c, \tag{C42}$$

$$\overline{\tau^2} = I_c^2 \log(1/c). \tag{C43}$$

Thus the moments of quantity τ are quite different from those of the interaction energy I.

D. The Zero-Frequency Limit of $K_i^I(z)$.

When evaluating on page 78 the singular part of the memory function, we replaced $K_i(t)$, (5.54), in an ad hoc fashion by a Debye relaxator, (5.60). Here we discuss $K_i(t)$ in more detail by factorizing three-spin correlation functions according to (5.61),

$$K_i^I(t) = \sum_{j,k \neq i} \frac{J_{ij}^2 J_{jk}^2}{2 I_i^2} \sum_{\pm} \langle \sigma_x^i(\pm t) \sigma_x^j(\pm t) \sigma_x^i \sigma_x^j \rangle \langle \sigma_z^k(\pm t) \sigma_z^k \rangle, \tag{D1}$$

the first of which accounts for correlations of the composite operators $\sigma_x^i \sigma_x^j$. When we define

$$C_{ij}(t) = (\sigma_x^i(t) \sigma_x^j(t) | \sigma_x^i \sigma_x^j) \tag{D2}$$

and proceed as above, we find the spectrum

$$K_i^{I\prime\prime}(\omega) = \sum_{j,k \neq i} \frac{J_{ij}^2 J_{jk}^2}{2 I_i^2} (C_{ij}(\omega) * G_k(\omega)), \tag{D3}$$

which requires knowledge of the spectral weight of the correlation $C_{ij}''(\omega)$ at zero frequency. Starting from

$$C_{ij}(z) = (\sigma_x^i \sigma_x^j [\mathcal{L} - z]^{-1} | \sigma_x^i \sigma_x^j), \tag{D4}$$

we construct a continued fraction similar to that for (5.12). Taking the time derivative of $A_1 = \sigma_x^i \sigma_x^j$ yields with the projection (5.14) $\tilde{A}_2 = \mathcal{Q}_1 \dot{A}_1$. When we use the decoupling approximation (5.46) and $(\tilde{A}_2 | \tilde{A}_2) = \tau_i^2 + \tau_j^2 \equiv \mathcal{N}^{-2}$, we get the normalized derivative

$$A_2 = -\mathcal{N} \sum_k{}' \left(J_{ik} \sigma_y^i \sigma_x^j \sigma_z^k + J_{jk} \sigma_x^i \sigma_y^j \sigma_z^k \right), \tag{D5}$$

where the prime at the sum indicates $i \neq k \neq j$ and where τ_i^2 and τ_j^2 are given by (5.47) with the term J_{ij}^2 suppressed. (Compare (C32) on p. 197.)

In the next step, we obtain with $\tilde{A}_3 = \mathcal{Q}_2 \dot{A}_2$ quite an involved expression,

$$\tilde{A}_3 = \mathcal{N} \Big[\Delta_0 \sum_k{}' \left(J_{ik} \sigma_x^j (\sigma_z^i \sigma_z^k - \sigma_y^i \sigma_y^k) + J_{jk} \sigma_x^i (\sigma_z^j \sigma_z^k - \sigma_y^j \sigma_y^k) \right)$$
$$+ \sum_{kl}{}'' \left(J_{ik} J_{il} + J_{jk} J_{jl} \right) \sigma_x^i \sigma_x^j \sigma_z^k \sigma_z^l - 2 \sum_{kl}{}' J_{ik} J_{jl} \sigma_y^i \sigma_y^j \sigma_z^k \sigma_z^l \Big], \tag{D6}$$

D. The Zero-Frequency Limit of $K_i^I(z)$.

where, as above, the prime confines the sums to $i \neq k \neq j$ and $i \neq l \neq j$, and the double prime adds the restriction $k \neq l$. With the memory function $M_{ij}(z) = (\tilde{A}_3|[\mathcal{Q}_2\mathcal{L} - z]^{-1}|\tilde{A}_3)$, we find for (D4)

$$C_{ij}(z) = \frac{-1}{z - \dfrac{\tau_i^2 + \tau_j^2}{z + M_{ij}(z)}}. \tag{D7}$$

To proceed further, we separate the first term in square brackets in (D6) from the remainder; when we perform normalization and drop corrections similar to $\tilde{\sigma}_x^j$ in (5.49), we have with (5.45)

$$A_3 = 2^{-1/2}\mathcal{N}\sum_k{}'\Big(J_{ik}\sigma_x^j(\sigma_z^i\sigma_z^k - \sigma_y^i\sigma_y^k) + J_{jk}\sigma_x^i(\sigma_z^j\sigma_z^k - \sigma_y^j\sigma_y^k)\Big). \tag{D8}$$

We take the time derivative once more and project on $\mathcal{Q}_3$,

$$\tilde{A}_4 = \sqrt{2}\mathcal{N}\Delta_0\sum_k{}'\Big(J_{ik}\sigma_x^j(\sigma_y^i\sigma_z^k + \sigma_z^i\sigma_y^k) + J_{jk}\sigma_x^i(\sigma_y^j\sigma_z^k - \sigma_z^j\sigma_y^k)\Big) + R_4, \tag{D9}$$

where R_4 contains terms involving products of four spin operators similar to those in (D6). When we define $N_{ij}(z) = (\tilde{A}_4|[\mathcal{Q}_3\mathcal{L} - z]^{-1}|\tilde{A}_4)$ and denote the memory functions arising from the second and third term in (D6) by $Y_3(z)$ and those from R_4 by $Y_4(z)$, we obtain the continued fraction

$$C_{ij}(z) = \frac{-1}{z - \dfrac{\tau_i^2 + \tau_j^2}{z - Y_3(z) + \dfrac{2\Delta_0^2}{z + N_{ij}(z) + Y_4(z)}}}. \tag{D10}$$

We are going to evaluate $C_{ij}(z \to 0)$ in an approximation similar to that of Sect. 5.1. First we discard composite operators made of four spins, i.e., we drop the memory functions Y_3 and Y_4. Second we evaluate the remaining function $N_{ij}(z)$ in Markov approximation. When we decouple $N_i(t)$ as in section 5.1 and take the Fourier transform, we obtain a convolution involving $\omega^2 G_k'''(\omega)$. Regarding the zero frequency limit, the factor ω^2 keeps $N_{ij}''(0) \equiv \eta$ small. With the definition

$$S_{ij} = \frac{2\Delta_0^2}{2\Delta_0^2 + \tau_i^2 + \tau_j^2}, \tag{D11}$$

we easily find

$$C_{ij}(z) = -S_{ij}\frac{1}{z + i(1 - S_{ij})\eta} \quad \text{for} \quad z \to 0. \tag{D12}$$

We will not attempt evaluation of the rate η; for our purpose only the spectral weight at zero frequency is significant, which is given by (D11).

Both function $C_{ij}(z)$ and $G_k(z)$ display a zero-frequency pole, thus giving rise to a relaxation contribution to $K_i^I(z)$, which we denote by $K_i(z)$,

D. The Zero-Frequency Limit of $K_i^I(z)$.

$$K_i(z) = -\sum_{j,k \neq j} \frac{J_{ij}^2 J_{jk}^2}{I_i^2} S_{ij} R_k \frac{1}{z + i\eta + i\gamma_k}. \tag{D13}$$

The corresponding spectral weight reads

$$\kappa_i^2 = \sum_{j,k \neq j} \frac{J_{ij}^2 J_{jk}^2}{I_i^2} S_{ij} R_k. \tag{D14}$$

We proceed as in Chap. 5 and retain only the contribution from the nearest neighbour i_0,

$$\kappa_i^2 = \frac{2\tau_{i_0}^2 \Delta_0^2}{\tau_i^2 + \tau_{i_0}^2 + 2\Delta_0^2} R_{i_0} \equiv \lambda_i \Delta_0^2; \tag{D15}$$

here we have defined the quantity λ_i which takes values between 0 and 2. Again, when we insert this in (5.69), we obtain an expression much too complicated to be averaged as it stands. In order to calculate the average relaxation amplitude, we replace all quantities in (D15) with their mean values and thus obtain

$$\lambda_0 = \frac{\log(1/c)\mu^2}{1 + \log(1/c)\mu^2} \overline{R}. \tag{D16}$$

After inserting this in (5.69) and using the distribution (C26), we find the self-consistency condition for $\overline{R}$

$$\overline{R} = \mu\sqrt{\lambda_0} \left[\frac{\pi}{2} - \arctan\left(\mu\sqrt{\lambda_0}\right) \right]. \tag{D17}$$

Comparison with the results obtained in Chap. 5, (5.27,5.113) reveals a similar functional form but with μ multiplied by a factor $\sqrt{\lambda_0}$. All three functions show a similar cross-over at $\mu \approx 1$, albeit with different power laws for the limiting cases $\mu \to 0$ and $\mu \to \infty$. Finally we remark that the decoupling scheme given in (5.62) has been studied in [60]; the resulting self-consistent equation for the relaxation amplitude yields a solution very similar to that given here.

E. Small-Polaron Approach

E.1 Time Correlation Functions of Polaron Operators

Here we derive relevant time correlation functions for the operators $B_\pm$; for a more complete treatment see the book of Mahan [163], or [88,180]. With (8.5) and (8.18) the time dependence is given by

$$B_\pm(t) = \prod_k \exp\left(\pm u_k(b_k e^{-i\omega_k t} - b_k^\dagger e^{i\omega_k t})\right); \tag{E1}$$

when we use the equality $e^{A+B} = e^A e^B e^{\frac{1}{2}[A,B]}$ repeatedly, we find

$$B_\pm(t) = \prod_k \exp\left(\mp u_k b_k^\dagger(t)\right) \exp\left(\pm u_k b_k(t)\right) \exp\left(-\tfrac{1}{2} u_k^2\right) \tag{E2}$$

$$B_\pm(t) B_\mp(t') = \prod_k \exp\left(\mp u_k b_k^\dagger \kappa_k(t, -t')\right) \exp\left(\pm u_k b_k \kappa_k(-t, t')\right)$$
$$\times \exp\left(-u_k^2 \kappa_k(0, t-t')\right), \tag{E3}$$

where we have defined

$$\kappa_k(t, t') = e^{i\omega_k t} - e^{-i\omega_k t'}. \tag{E4}$$

The thermal average is to be taken with respect to $e^{-\beta H_B}$; when we note $\langle e^{zb_k^\dagger} e^{-z^* b_k}\rangle = e^{-|z|^2 n_k}$, and use $\kappa_k(t,t') = \kappa_k^*(-t,-t')$ and the phase definition (8.54), we obtain

$$\langle B_\pm \rangle = \exp\left[-\Sigma_k u_k^2 (n_k + \tfrac{1}{2})\right], \tag{E5}$$

$$\langle B_\pm(t) B_\mp(t')\rangle = \langle B_\pm\rangle^2 e^{\varphi(t-t')}. \tag{E6}$$

Following the same argument as above, we find

$$\langle B_\pm(t) B_\pm(t')\rangle = \langle B_\pm\rangle^2 e^{-\varphi(t-t')}. \tag{E7}$$

Evaluation of the four-time correlation functions is rather more tedious; we note the intermediate formulae

$$B_\pm(t) B_\mp(\tau) B_\pm(\tau') B_\mp(t')$$
$$= \prod_k \exp\left[\pm u_k b_k^\dagger (\kappa_k(t,\tau) + \kappa_k(\tau',t'))\right]$$
$$\times \exp\left[\mp u_k b_k (\kappa_k^*(t,\tau) + \kappa_k^*(\tau',t'))\right] f_+(t,\tau,\tau',t'), \tag{E8}$$

204   E. Small-Polaron Approach

$$B_\pm(t)B_\mp(\tau)B_\mp(\tau')B_\pm(t')$$
$$= \prod_k \exp\left[\pm u_k b_k^\dagger(\kappa_k(t,\tau) - \kappa_k(\tau',t'))\right]$$
$$\times \ \exp\left[\mp u_k b_k(\kappa_k^*(t,\tau) - \kappa_k^*(\tau',t'))\right] f_-(t,\tau,\tau',t'), \quad \text{(E9)}$$

where κ^* is the complex conjugate of κ and where we have defined

$$f_\pm(t,\tau,\tau',t') = \exp\left[-u_k^2 \kappa_k(0,t-\tau) - u_k^2 \kappa_k(0,\tau'-t')\right.$$
$$\left.\mp u_k^2 \kappa_k(-t,\tau)\kappa_k(\tau',-t')\right]. \quad \text{(E10)}$$

Equations (E8–E10) finally yield with (8.14) and (5.46)

$$\langle K(t,\tau)K(\tau',t')\rangle = \frac{\tilde{\Delta}_0^4}{\hbar^4} e^{\varphi(t-\tau)+\varphi(\tau'-t')} \cosh[\Phi(t,\tau,\tau',t')], \quad \text{(E11)}$$

where we have defined the function

$$\Phi(t,\tau,\tau',t') = \varphi(t-t') + \varphi(\tau-\tau') - \varphi(t-\tau') - \varphi(\tau-t'). \quad \text{(E12)}$$

According to the above process, correlations of any number of operators $B_\pm$ can be calculated. We quote the general expression from [180,88]

$$\langle B_{\alpha_1}(\tau_1)B_{\alpha_2}(\tau_2)...B_{\alpha_N}(\tau_N)\rangle$$
$$= \langle B_\pm\rangle^N \exp\left[-\Sigma_{i<j}\alpha_i\alpha_j\varphi(\tau_i-\tau_j)\right], \quad \text{(E13)}$$

with $\alpha_i = \pm 1$ and the sums in the exponential running from 1 to N.

E.2 Upper and Lower Bounds for Γ_1

For convenience we note the integrals

$$\int_0^\infty dx \frac{x}{\sinh(x)} = \frac{\pi^2}{4} \equiv J_1, \quad \text{(E14)}$$

$$\int_0^\infty dx \frac{x^2}{\sinh(x)^2} = \frac{\pi^2}{6} \equiv J_2, \quad \text{(E15)}$$

$$\int_0^\infty dx \frac{x}{e^x - 1} = \frac{\pi^2}{6}, \quad \text{(E16)}$$

which have been repeatedly referred to in Chap. 8.

Upper Bound $\mathcal{O}$

According to (5.51), we have $c(x) \leq 1$; thus the sum in (8.78) fulfils the inequality

$$\sum_{\{\pm\}} c(\pm x_1 ... \pm x_{n-1}) \leq 2^{n-1} \quad \text{(E17)}$$

for all values of $x_1, ..., x_{n-1}$. When we insert this in (8.78) and perform the remaining integrations, we find with (E14)

$$a_n \le (2J_1)^{n-1}. \tag{E18}$$

With (E18) in (8.79), it is straightforward to derive an upper bound for the damping rate

$$\mathcal{O} = \frac{\tilde{\Delta}_0^2}{2\hbar k_B T} \frac{\pi}{2J_1} \left(e^{8J_1 \tilde{\alpha} T^2} - 1 \right). \tag{E19}$$

Lower Bound $\mathcal{U}$

Now we are going to derive a lower bound for the coefficients a_n. When we take $y = |\pm x_1 ... \pm x_{n-2}|$ and $x = x_{n-1}$ for $n > 2$, we have

$$\sum_{\{\pm\}} c(\pm x_1 ... \pm x_{n-1}) = \sum_{\{\pm\}} c(y-x) + c(y+x), \tag{E20}$$

where the sums run over $n-1$ signs on the left-hand side and over $n-2$ signs on the right-hand side. Since $c(x) \le 1$, and $c(x)$ is a monotonously decreasing function, the first term on the right-hand side of (E20) fulfils the inequality

$$c(y-x) > c(y)c(x). \tag{E21}$$

[Note $y \ge 0$, $x \ge 0$, and $c(x) = c(-x)$.]

Regarding the second term on the right-hand side of (E20), we deal separately with the cases $y > x$ and $y < x$. For the first case we find with $y+x > y$ and $\sinh(x+y) < 2\sinh(y)\cosh(x)$ the relation

$$c(y+x) > c(y) \frac{1}{2\cosh(x)} \quad \text{for } y > x; \tag{E22}$$

for the second one we note the obvious inequality

$$c(y+x) > c(y)c(2x) \quad \text{for } y < x. \tag{E23}$$

When we use $c(2x) = c(x)\cosh(x)^{-1}$ and $\cosh(x) \ge 1$, we finally obtain

$$c(y-x) + c(y+x) > c(y)c(x) \left(1 + \frac{1}{2\cosh(x)^2} \right). \tag{E24}$$

Taking $y = |\pm x_1 ... \pm x_{n-3}|$ and $x = x_{n-2}$, and repeating the above argument, we find the sum in (8.78) to fulfil the inequality

$$\sum_{\{\pm\}} c(\pm x_1 \pm ... \pm x_{n-1}) \ge \prod_{j=1}^{n-1} c(x_j) \left(1 + \frac{1}{2\cosh(x_j)^2} \right) \tag{E25}$$

for $n > 2$. After we have inserted this in (8.78), the $(n-1)$ integrations become independent of each other; with $c(x)^2 \cosh(x)^{-2} = c(2x)^2$ and the integral (E15) we find

E. Small-Polaron Approach

$$\int_0^\infty dx \left(c(x)^2 + \frac{1}{2} c(2x)^2 \right) = \left(1 + \frac{1}{4} \right) J_2,$$

which yields the inequality

$$a_n > [(1 + \tfrac{1}{4}) J_2]^{n-1}. \tag{E26}$$

Equation (E26) holds true for $n = 1$ and $n > 2$; for the case $n = 2$, we note the value $a_2 = J_2 = \frac{\pi^2}{6}$; with (8.79) we derive the lower bound

$$\mathcal{U} = \frac{\tilde{\Delta}_0^2}{2\hbar k_B T} \frac{\pi}{\frac{5}{4} J_2} \left(e^{\frac{5}{4} J_2 4 \tilde{\alpha} T^2} - 1 \right). \tag{E27}$$

Equations (E19,E27) yield rigorous bounds for the damping rate Γ_1,

$$\mathcal{U} < \Gamma_1 < \mathcal{O}. \tag{E28}$$

These bounds depend on temperature, both through the exponential factors and through the tunnel energy $\tilde{\Delta}_0$. In order to render the temperature variation more obvious, we rewrite (E19,E27) using $J_2 = \frac{\pi^2}{6}$ and the constant (8.50),

$$\mathcal{O} = \frac{\Delta_0^2}{\pi \hbar k_B T} \left[\exp\left(\tfrac{4}{3} \pi^2 \tilde{\alpha} T^2 \right) - \exp\left(-\tfrac{2}{3} \pi^2 \tilde{\alpha} T^2 \right) \right] \tag{E29}$$

$$\mathcal{U} = \frac{12}{5} \frac{\Delta_0^2}{\pi \hbar k_B T} \left[\exp\left(\tfrac{1}{6} \pi^2 \tilde{\alpha} T^2 \right) - \exp\left(-\tfrac{2}{3} \pi^2 \tilde{\alpha} T^2 \right) \right]. \tag{E30}$$

For weak coupling or low temperatures, both $\mathcal{U}$ and $\mathcal{O}$ tend towards the one-phonon rate,

$$\mathcal{U} = \mathcal{O} = (2\pi/\hbar k_B) \tilde{\alpha} \Delta_0^2 T \quad \text{for } \tilde{\alpha} T^2 \ll 1, \tag{E31}$$

whereas at the opposite limit both increase exponentially.

References

1. F. Hund: Z. Phys. B **43**, 805 (1927)
2. E. F. Barker: Phys. Rev. **33**, 684 (1929)
3. C. E. Cleeton, N. H. Williams: Phys. Rev. **45**, 234 (1934)
4. A. O. Caldeira, A. J. Leggett: Phys. Rev. Lett. **46**, 211 (1981); Ann. Phys. (N.Y.) **149**, 374 (1983) and **153**, 445 (1983)
5. J. S. Langer, V. Ambegaokar: Phys. Rev. **164**, 498 (1967)
6. J. Kondo: Physica **84** B, 40 (1976); **84** B, 207 (1976); **125** B, 279 (1984); **126** B, 377 (1984)
7. H. Frauenfelder, P. G. Wolynes: Science **229**, 337 (1985)
8. J. S. Langer: Ann. Phys. (N. Y.) **41**, 108 (1967)
9. S. Chakravarty: Phys. Rev. Lett. **49**, 681 (1982)
10. P. W. Anderson, G. Yuval, D. R. Hamann: Phys. Rev. B **1**, 4464 (1970); P. W. Anderson, G. Yuval: J. Phys. C **4**, 607 (1971)
11. A. J. Bray, M. A. Moore: Phys. Rev. Lett. **49**, 1545 (1982)
12. S. Chakravarty, A. J. Leggett: Phys. Rev. Lett. **52**, 5 (1984)
13. H. Grabert, U. Weiss: Phys. Rev. Lett. **54**, 1605 (1985)
14. M. P. A. Fisher, A. T. Dorsey: Phys. Rev. Lett. **54**, 1609 (1985)
15. A. J. Leggett, S. Chakravarty, A. T. Dorsey, M. P. A. Fisher, A. Garg, W. Zwerger: Rev. Mod. Phys. **59**, 1 (1987)
16. U. Weiss, H. Grabert, S. Linkwitz: J. Low. Temp. Phys. **68**, 213 (1987)
17. U. Weiss, H. Grabert, P. Hänggi, P. Riseborough: Phys. Rev. B **35**, 9535 (1987)
18. U. Weiss, M. Wollensack: Phys. Rev. Lett. **62**, 1663 (1989)
19. H. Grabert, S. Dattagupta, R. Jung: J. Phys. C **1**, 1405 (1989)
20. P. Hänggi, P. Talkner, M. Borkovec: Rev. Mod. Phys. **62**, 251 (1990)
21. U. Weiss: Quantum Dissipative Systems, Series in Modern Condensed Matter Physics, Vol. 2. World Scientific, Singapore (1993)
22. S. Chakravarty, J. Rudnick: Phys. Rev. Lett. **75**, 501 (1995)
23. H. Grabert, H. R. Schober: Hydrogen in Metals III, H. Wipf (Ed.), Springer Heidelberg (1996)
24. W. Zwerger: Z. Phys. B **53**, 53 (1983); ibid **54**, 87 (1983)
25. W. Götze, G. M. Vujicic: Phys. Rev. B **38**, 9398 (1988)
26. G. Weiss, W. Arnold, K. Dransfeld, H. J. Güntherodt: Solid State Comm. **33**, 111 (1980)
27. D. Steinbinder, H. Wipf, H. R. Schober, H. Blank, G. Kearley, C. Vettier, A. Magerl: Europhys. Lett. **8**, 269 (1989)
28. N.E. Byer, H.S. Sack: Phys. Rev. Lett. **17**, 72 (1966)
29. N.E. Byer, H.S. Sack: J. Phys. Chem. Solids **29**, 677 (1968)
30. V. Narayanamurti, R.O. Pohl: Rev. Mod. Phys. **42**, 201 (1970)
31. F. Bridges: Crit. Rev. Solid State Sci. **5**, 1 (1975); F. Holuj, F. Bridges: Phys. Rev. B **27**, 5286 (1983)

32. M. Gomez, S.P. Bowen, J.A. Krumhansl: Phys. Rev. **153**, 1009 (1967)
33. G. Leibfried, N. Breuer: Point Defects in Metals, Springer Tracts in Modern Physics Vol. 81, Springer Berlin Heidelberg New York 1978.
34. E. Kanda, T. Goto, H. Yamada, S. Suto, S. Tanaka, T. Fujita, T. Fujimura: J. Phys. Soc. Jap. **54**, 175 (1985)
35. J. P. Sethna: Phys. Rev. B **24**, 698 (1981) and **25**, 5050 (1982)
36. A. Würger: Z. Phys. B **76**, 65 (1989)
37. A.T. Fiory: Phys. Rev. B **4**, 614 (1971); Rev. Sci. Instr. **42**, 930 (1971)
38. K. Knop, W. Känzig: Phys. Kondens. Materie **15**, 205 (1972); Helvetica Physica Acta **46**, 889 (1974)
39. R.C. Potter, A.C. Anderson: Phys. Rev. B **24**, 677 (1981)
40. R. Weis, C. Enss, B. Leinböck, G. Weiss, S. Hunklinger: Phys. Rev. Lett. **75**, 2220 (1995)
41. R. Weis: Doktorarbeit Heidelberg (1995), unpublished
42. R. Weis, C. Enss, A. Würger, F. Lüty: submitted to Phys. Rev. B.
43. M. Hübner: Diplomarbeit Heidelberg (1995), unpublished; M. Hübner, C. Enss, G. Weiss: to be published
44. R. Kubo, M. Toda, N. Hashitsume: Statistical Physics, 2. Non-equilibrium Statistical Mechanics (Springer series in solid state sciences Vol. 30). Springer Berlin Heidelberg New York 1985.
45. J.N. Dobbs, A.C. Anderson: Phys. Rev. B **33**, 4172 (1986)
46. W.N. Lawless: Phys. Kond. Mat. **5**, 100 (1966)
47. M.W. Klein: Phys. Rev. **141**, 489 (1966)
48. M.E. Baur, W.R. Salzman: Phys. Rev. **178**, 1440 (1969)
49. M.W. Klein: Phys. Rev. B **29**, 5825 (1984); Phys. Rev. B **31**, 2528 (1985); Phys. Rev. B **40**, 1918 (1989)
50. T. Kranjc: J. Phys. A **25**, 3065 (1992)
51. O. Terzidis, A. Würger: Z. f. Phys. B **94**, 341 (1994)
52. O. Terzidis: Doktorarbeit Heidelberg (1995), unpublished
53. O. Terzidis, A. Würger: J. Phys. Cond. Matt. **8**, 7303 (1996)
54. P. P. Peressini, J. P. Harrison, R. O. Pohl: Phys. Rev. **180**, 939 (1969) and **182**, 926 (1969)
55. D. Walton: Phys. Rev. Lett. **19**, 305 (1967)
56. W. D. Seward, V Narayanumurti: Phys. Rev. **148**, 463 (1966)
57. J. P. Harrison, P. P. Peressini, R. O. Pohl: Phys. Rev. **171**, 1037 (1968)
58. R. F. Wielinga, A. R. Miedema, W. J. Huiskamp: Physica **32**, 1568 (1966)
59. A. Würger: Z. f. Phys. B **94**, 173 (1994)
60. A. Würger: Z. f. Phys. B **98**, 561 (1995)
61. A. Würger: unpublished
62. C. Enss, M. Gaukler, M. Tornow, R. Weis, A. Würger: Phys. Rev. B **53**, 12094 (1996)
63. A. Würger, R. Weis, M. Gaukler, C. Enss: Europhys. Lett. **33**, 533 (1996)
64. R. Weis, C. Enss: to appear in: Proceedings of the Conference on Low-Temperature Physics 1996, Czechoslowak Journal of Physics (1996)
65. C. Enss, M. Gaukler, M. Nullmeier, R. Weis, A. Würger: to be published
66. F. G. Fumi, M. P. Tosi: J. Phys. Chem. Solids **25**, 31 (1964)
67. M. C. Robinson, A. C. Hollis Hallett: Canad. J. Phys. **44**, 2211 (1966)
68. F, Holuj, F. Bridges: Phys. Rev. B **27**, 5286 (1983)
69. X. Wang, F. Bridges: Phys. Rev. B **46**, 5122 (1992)
70. M. C. Hetzler, D. Walton: Phys. Rev. Lett. **24**, 505 (1970); Phys. Rev. B **8**, 4801 and 4812 (1973)
71. H. S. Sack, M. C. Moriarty: Solid State Comm. **3**, 93 (1965)
72. D. Moy, R. C. Potter, A. C. Anderson: J. of Low-Temp. **52**, 399 (1983)

73. S. Kapphan, F. Lüty: J. Phys. Chem. Solids **34**, 969 (1973)
74. H. B. Shore, L. M. Sander: Phys. Rev. B **12**, 1546 (1975)
75. W. Känzig, H. R. Hart, S. Roberts: Phys. Rev. Lett. **13**, 543 (1964)
76. R. P. Lowndes, D. H. Martin: Proc. Royal Soc. A **316**, 351 (1970)
77. R. Windheim, H. Kinder: Phys. Lett. **51 A**, 475 (1975)
78. S. Kapphan: J. Phys. Chem. Solids **35**, 621 (1974)
79. S. Haussühl: Z. Naturforsch. **12a**, 445 (1957)
80. L. D. Landau, E. M. Lifshitz: Quantum mechanics, Pergamon Press Oxford (1977)
81. A. Messiah: Quantum mechanics Vol. I, North-Holland Amsterdam (1985)
82. H. Horner: private communication (1995)
83. N. C. Wong, S. S. Kano, R. C. Brewer: Phys. Rev. A **21**, 260 (1980)
84. G. Baier, M. von Schickfus, C. Enss: Europhys. Lett. **8**, 487 (1989); G. Baier, C. Enss, M. von Schickfus: Phys. Rev. B **40**, 9868 (1989)
85. A. Würger: Z. f. Phys. B **93**, 109 (1993)
86. T. Holstein: Ann. Phys. (N.Y.) **8**, 343, (1959)
87. C. P. Flynn, A. M. Stoneham: Phys. Rev. B **1**, 3966 (1970)
88. R. Pirc, P. Gosar: Phys. Kond. Mat. **9**, 377 (1969)
89. R. Beck, W. Götze, P. Prelovsek: Phys. Rev A **20**, 1140 (1979)
90. Yu. Kagan, L. A. Maksimov,: Sov. Phys. JETP **38**, 307 (1974)
91. H. Grabert, U. Weiss, H. R. Schober: Hyperfine Intact. **31**, 147 (1986); H. Grabert: Phys. Rev. B **46**, 12753 (1992)
92. H. Teichler, A. Seeger: Phys. Lett. **82**, 91 (1981)
93. R. Silbey, R. A. Harris: J. Phys. Chem. **93**, 7062 (1989)
94. P. Neu: Doktorarbeit Universität Heidelberg (1994), unpublished
95. P. Neu, A. Würger: Z. Phys. B **95**, 385 (1994)
96. P. Neu, A. Würger: Europhysics Letters **27**, 457 (1994)
97. A. Würger: Habilitationsschrift Heidelberg (1995), unpublished
98. A. Würger: Z. Phys. B **85**, 93 (1991)
99. M. Tornow, R. Weis, G. Weiss, C. Enss, S. Hunklinger: Physica B **194-196** 1063 (1994); M. Tornow: Diplomarbeit Heidelberg (1993), unpublished
100. H. B. Rosenstock: J. Phys. Chem. Sol. **23**, 659 (1962); J. Non-Cryst. Sol. **7**, 351 (1972)
101. H. P. Baltes: Solid State Commun. **13**, 225 (1973)
102. P. Fulde, F. Wagner: Phys. Rev. Lett. **27**, 1280 (1971)
103. P. W. Anderson, B. I. Halperin, C. Varma: Philos. Mag. **25**, 1 (1972)
104. W. A. Phillips: J. Low. Temp. Phys. **7**, 351 (1972)
105. J. Jäckle: Z. Phys. **257**, 212 (1972)
106. R. C. Zeller, R. O. Pohl: Phys. Rev B **4**, 2029 (1971)
107. S. Hunklinger, W. Arnold: In: Thurston R. N., Mason W. P. (Eds.): Physical Acoustics, Vol. 12, Academic Press, New York (1976)
108. J. C. Lasjaunias, A. Ravex, M. Vandorpe, S. Hunklinger: Sol. State Commun. **17**, 1045 (1975)
109. S. Hunklinger, A. K. Raychaudhuri: In: D. F. Brewer (Ed.): Progress in low temperature physics, Vol. IX, Elsevier, Amsterdam (1986)
110. L. Piché, R. Maynard, S. Hunklinger, J. Jäckle: Phys. Rev. Lett. **32**, 1426 (1974)
111. J. Jäckle, L. Piché, W. Arnold, S. Hunklinger: J. Non-Cryst. Solids **20**, 365 (1976)
112. The discussion given in Europhys. Lett. **28**, 597 (1994) by the present author is incorrect because of an improper choice of the dependence of the double-well potential on the sound wave amplitude; this cancels in particular the linear temperature dependence of the damping function derived in that paper.

113. Encyclopaedia Britannica, 15. edition, Vol. 8, p. 209. Benton Chicago 1976
114. J. Classen, C. Enss, C. Bechinger, G. Weiss, S. Hunklinger: Ann. Physik (Leipzig) **3**, 315 (1994)
115. J. L. Black, P. Fulde: Phys. Rev. Lett. **43**, 453 (1979)
116. C. Enss: Dissertation Heidelberg 1990, unpublished
117. C. Enss, H. Schwoerer, D. Arndt, M. von Schickfus, G. Weiss: Europhys. Lett. **26**, 289 (1994)
118. R. Weis: Diplomarbeit Heidelberg (1992), unpublished
119. H. Mori: Progr. Theor. Phys. **33**, 127 (1965); ibid. **34**, 399 (1965)
120. R. Zwanzig: J. Chem. Phys. **33**, 1338 (1960); Physica **30**, 1109 (1964)
121. S. F. Edwards, P. W. Anderson: J. of Phys. F **5**, 965 1975)
122. W. Götze, L. Sjögren: J. Phys. C **17**, 5759 (1984) Z. Phys. B **65**, 415 (1987)
123. V. L. Aksenov, M. Bobeth, N. M. Plakida, J. Schreiber: J. Phys. C **20**, 375 (1987)
124. Compared to [96] we have changed the definition of $\tilde{\gamma}$. To get the expression used here, $\tilde{\gamma}$ defined in [96] must be multiplied by $\pi/\hbar$.
125. R.E. Strakna, H.T. Savage: J. Appl. Phys. **35**, 1445 (1964)
126. M. G. R. Zobel: Thesis, cited after ref. [127]
127. S. Rau, PhD-thesis Universität Heidelberg (1995), unpublished
128. S. Rau, C. Enss, S. Hunklinger, P. Neu, A. Würger: Phys. Rev. B **52**, 7179-7195 (1995)
129. C.R. Kurkjian, J.T. Krause: J. Am. Ceram. Soc. **49**, 171 (1969)
130. P. J. Bray: Interaction of Radiation with Solids, Plenum New York (1970)
131. P. Neu, A. Würger: Europhys. Lett. **29** 561 (1995)
132. W. Brenig: Statistical Theory of Heat: Non-equilibrium Phenomena, Springer Berlin Heidelberg New York (1989)
133. W. A. Phillips (Ed.): Amorphous Solids – Low-Temperature Properties, Springer, Berlin Heidelberg New York (1981)
134. S. Rau, S. Hunklinger: Private communication (1995)
135. J. T. Krause: Phys. Lett. **43** A, 325 (1973)
136. G. Bellessa: Phys. Rev. Lett. **40**, 1456 (1978)
137. G. Bellessa, C. Lemercier, D. Caldemaison: Phys. Lett. **62** A, 127 (1977)
138. G. Bellessa, P. Doussineau, A. Levelut: J. Phys. (Paris) Lett. **38**, L65 (1977)
139. P. J. Anthony, A. C. Anderson: Phys. Rev. B **20**, 763 (1979)
140. P. W. Anderson: J. Phys. (Paris) Colloq. **37** C4-339 (1976)
141. P. Doussineau, C. Frenois, R. G. Leisure, A. Levelut, J.-Y. Prieur: J. Physique **41**, 1193 (1980)
142. C. C. Yu, A. J. Leggett: Comments Cond. Matt. Phys. **14**, 231 (1988)
143. D. Tielbürger, R. Merz, R. Ehrenfels, S. Hunklinger: Phys. Rev. B **45**, 2750 (1992)
144. U. Buchenau, Yu. M. Galperin, V. L. Gurevich, D. A. Parshin, M. A. Ramos, H. R. Schober: Phys. Rev. B **46**, 2798 (1992)
145. U. Buchenau, Yu. M. Galperin, V. L. Gurevich, H. R. Schober: Phys. Rev. B **43**, 5039 (1991)
146. D. A. Parshin: Physica Scripta **T49**, 180 (1993)
147. F. N. Ignatiev, V. G. Karpov, M. Klinger: J. Non-Cryst. Sol. **55**, 307 (1983)
148. A. L. Burin, Yu. Kagan: Physica B **194-196**, 393 (1994)
149. S. Hunklinger, W. Arnold, S. Stein, R. Nava, K. Dransfeld: Phys. Lett. A **42**, 253 (1972)
150. A. C Anderson: In: W. A. Phillips (Ed.): Amorphous Solids, Topics in Current Physics Vol. 24, Springer Berlin Heidelberg New York (1981)
151. M. von Schickfus, S. Hunklinger: Phys. Lett. A **64**, 144 (1977)
152. P. Esquinazi, R. König, F. Pobell: Z. Phys. B **87**, 305 (1992)

153. P. Esquinazi, R. König, D. Valentin, F. Pobell: J. of All. and Comp. **211**, 27 (1994)
154. R. König, P. Esquinazi, B. Neppert: Phys. Rev. B **51** 11424 (1995)
155. E. Gaganidze, P. Esquinazi, R. König: Europhys. Lett. **31** 13 (1995)
156. B. E. White Jr., R. O. Pohl: Z. Phys. B **100**, 401 (1996)
157. A. Nittke, M. Scherl, P. Esquinazi, W. Lorenz, Junyun Li, F. Pobell: J. of Low-Temp. Phys. **98**, 517 (1995)
158. S. Sahling, A. Sahling, M. Kolac: Sol. State Commun. **65**,1031 (1988); D. A. Parshin, S. Sahling: Phys. Rev. B **47**, 5677 (1993)
159. G. Federle, S. Hunklinger: J. de Physique C **9**, Vol. 43, 505 (1982)
160. H. A. Kramers, Physica **7**, 284 (1940)
161. F. Haake: Statistical Treatment of Open Systems by Generalized Master Equations, Springer Tracts in Modern Physics Vol. 66, Springer Berlin Heidelberg New York (1973)
162. H. Grabert: Projection Operator Techniques in Non-equilibrium Statistical Mechanics, Springer Tracts in Modern Physics Vol. 95, Springer Berlin Heidelberg New York (1982)
163. G. D. Mahan: Many-particle physics, Plenum Press New York 1981
164. B. Daeubler, H. Risken, L. Schoendorff: Phys. Rev. B **46** 1654 (1992)
165. F. Bloch: Phys. Rev. **70** 460 (1946)
166. H. J. Carmichael, D. F. Walls: J. Phys. B **9**, 1199 (1976)
167. Yu. M. Galperin, V. L. Gurevich, D. A. Parshin: Phys. Rev B **37**, 10339 (1988)
168. J. R. Klauder, P. W. Anderson: Phys. Rev **125**, 912 (1962)
169. I.I. Rabi: Phys. Rev. **51**, 652 (1937)
170. J. M. Rowe, J. J. Rush, S. M. Shapiro, D. G. Hinks, S. Susman: Phys. Rev. B **21**, 4863 (1980)
171. R. F. Wood, M. Mostoller: Phys. Rev. Lett. **39**, 819 (1977)
172. S. Grossmann: Phys. Rev. A **17**, 1123 (1978)
173. R. Kubo: J. Phys. Soc. Japan **12** 570 (1957)
174. D. Forster: Hydrodynamic Fluctuations, Broken Symmetry and Correlation Functions, Benjamin Cummings, Reading, Mass. (1975)
175. E. Fick, G. Sauermann: Quantenstatistik dynamischer Prozesse, Bd.IIa, Verlag Harry Deutsch, (1986)
176. I. S. Gradstein, I. M. Ryshik: Summen-, Produkt- und Integraltafeln, Thun Frankfurt 1981
177. W. Götze: Solid State Comm. **27**, 1393 (1978)
178. W. Götze: In: J. P. Hansen et al. (Eds.): Liquids, freezing and the glass transition, Amsterdam, North-Holland (1989)
179. K. H. Michel: Phys. Rev. Lett. **57**, 460 (1986); C. Bostoen, K. H. Michel: Z. Phys. B **71**, 369 (1988)
180. I. G. Lang, Yu. A. Firsov: JETP **16**, 1301 (1963)

Index

Activated relaxation
- of interacting defects, 100
- of tunneling systems in glasses, 136, 149, 150

Asymmetry energy
- effect on rotary echoes, 42
- for a defect pair, 32, 47
- of two-level systems in glasses, 121

Average relaxation amplitude, 72, 97

Break-down of two-level description, 135, 149

Correlation spectrum, 186
- for interacting defects, 80
- for the spin-boson model, 128, 134

Coupling to elastic waves
- of substitutional defects, 18
- of tunneling systems in glasses, 124

Cross-over to relaxation
- driven by interaction, 109
- driven by temperature, mode-coupling theory, 129
- driven by temperature, small-polaron approach, 179
- of tunneling in glasses, 150

Damping
- due to dipolar interaction, 81
- of lithium tunneling, 18

Damping through sound waves
- of substitutional defects, 18
- of TLS in glasses, 128

Debye-Waller factor, 161, 180

Decoupling of memory functions, 185
- for interacting defects, 78
- for the spin-boson model, 132

Degenerate states
- in glasses, 119
- of substitutional defects, 10

Density of states
- of interacting defects, 110
- of interacting impurities, 84
- of the pair model, 57

Dielectric susceptibility
- classical limit, 112
- for the pair model, 64
- of a two-state system, 52
- of interacting impurities, 72, 92, 96
- of non-interacting impurities, 15
- of two coupled impurities, 35, 37, 38

Dipolar interaction
- of two impurities, 28
- of two-state systems, 53

Dipole moment
- of a single impurity, 14
- of a two-state system, 52

Elastic quadrupole moment, 16

Elastic response function
- in glasses, 137
- of a lithium impurity, 22

Energy eigenstates
- of a single impurity, 12
- of two coupled impurities, 34

Energy spectrum
- of the pair model, 55
- of two-level systems in glasses, 121

Equation of motion
- for a two-level system, 125, 156
- for interacting two-level systems, 69

Fermi's Golden Rule, 18
Fluctuation operators, 126
Fluctuation-dissipation theorem, 186
Fourier transformation, 185

Ground-state splitting
- of a single impurity, 12
- of two coupled impurities, 34, 38, 39

Internal energy
- of two coupled impurities, 35
Internal friction
- in glasses, 137
- in oxide glasses, 143
- of polystyrene, 148
- relaxational contribution, 139
- resonant contribution, 138
- thermally activated relaxation, 140
Isotope effect
- for relaxation amplitude, 100
- for relaxation rate, 99, 105
- on resonant absorption, 94
- on rotary echoes, 44, 48

Kanzani forces, 16
Kramers-Kronig relation, 185

Laplace transformation, 185
Lattice distortion, 16

Memory function
- for interacting defects, 78
- for the spin-boson model, 127, 130
Mode-coupling equations
- for interacting impurities, 106
- for the spin-boson model, 132

Ohmic damping, 123
Oxide glasses
- sound propagation, 143

Perturbation series
- for the damping rate, 163
Perturbation theory
- Rayleigh-Schrödinger, 29, 46
Polaron transformation, 154
Polycrystalline metals
- sound propagation, 146
Polymers
- sound propagation, 148

Rabi frequency, 41
Relaxation
- comparison with spin glasses, 106
- of TLS in glasses, 134, 149
Relaxation amplitude
- for interacting defects, 96

Relaxation peak
- for interacting defects, 101
- for tunneling in glasses, 141
Relevant operators, 107
Resolvent identity, 70

Saddle-point integration, 166
Schottky anomaly, 13, 88
Selection rules
- dielectric, 14
- elastic, 17
Self-consistent equation
- for relaxation amplitude, 201
- for the damping rate, 133
Soft-potential model, 136
Sound velocity
- in amorphous solids, 137
- in oxide glasses, 145
- in polycrystalline metals, 147
- of doped alkali halides, 22
- relaxational contribution, 139
- resonant contribution, 138
- thermally activated relaxation, 141
Specific heat
- for the pair model, 58
- of a single impurity, 13
- of glasses, 119
- of interacting impurities, 87
Spin operators
- for a lithium impurity, 51
Spin operators for TLS
- energy eigenstates, 126
- localized states, 120
Spin-boson model, 123, 153
Statistical operator
- for the pair model, 56
- of a two-level system in glasses, 121
Substitutional defect
- cyanide, 14
- lithium, 9

Thermal conductivity
- of doped alkali halides, 25
- of glasses, 116
Tunneling amplitude
- of a single impurity, 10
Two-state approximation
- for a defect pair, 40, 46, 54
- for a lithium impurity, 51

Springer Tracts in Modern Physics

120 Nuclear Pion Photoproduction
By A. Nagl, V. Devanathan, and H. Überall
1991. 53 figs. VIII, 174 pages

121 Current-Induced Nonequilibrium Phenomena
in Quasi-One-Dimensional Superconductors
By R. Tidecks 1990. 109 figs. IX, 341 pages

122 Particle Induced Electron Emission I
With contributions by M. Rösler, W. Brauer,
and J. Devooght, J.-C. Dehaes, A. Dubus, M. Cailler, J.-P. Ganachaud
1991. 64 figs. X, 130 pages

123 Particle Induced Electron Emission II
With contributions by D. Hasselkamp and H. Rothard, K.-O. Groeneveld,
J. Kemmler and P. Varga, H. Winter 1992. 90 figs. IX, 220 pages

124 Ionization Measurements in High Energy Physics
By B. Sitar, G. I. Merson, V. A. Chechin, and Yu. A. Budagov
1993. 184 figs. X, 337 pages

125 Inelastic Scattering of X-Rays with Very High Energy Resolution
By E. Burkel 1991. 70 figs. XV, 112 pages

126 Critical Phenomena at Surfaces and Interfaces
Evanescent X-Ray and Neutron Scattering
By H. Dosch 1992. 69 figs. X, 145 pages

127 Critical Behavior at Surfaces and Interfaces
Roughening and Wetting Phenomena
By R. Lipowsky 1996. 80 figs. X, Approx. 180 pages

128 Surface Scattering Experiments with Conduction Electrons
By D. Schumacher 1993. 55 figs. IX, 95 pages

129 Dynamics of Topological Magnetic Solitons
By V. G. Bar'yakhtar, M. V. Chetkin, B. A. Ivanov, and S. N. Gadetskii
1994. 78 figs. VIII, 179 pages

130 Time-Resolved Light Scattering from Excitons
By H. Stolz 1994. 87 figs. XI, 210 pages

131 Ultrathin Metal Films
Magnetic and Structural Properties
By M. Wuttig 1998. 103 figs. X, Approx. 180 pages

132 Interaction of Hydrogen Isotopes with Transition-Metals and Intermetallic Compounds
By B. M. Andreev, E. P. Magomedbekov, G. Sicking 1996. 72 figs. VIII, 168 pages

133 Matter at High Densities in Astrophysics
Compact Stars and the Equation of State
In Honor of Friedrich Hund's 100th Birthday
By H. Riffert, H. Müther, H. Herold, and H. Ruder 1996. 86 figs. XIV, 278 pages

134 Fermi Surfaces of Low-Dimensional Organic Metals and Superconductors
By J. Wosnitza 1996. 88 figs. VIII, 172 pages

135 From Coherent Tunneling to Relaxation
Dissipative Quantum Dynamics of Interacting Defects
By A. Würger 1996. 51 figs. VIII, 216 pages

136 Optical Properties of Semiconductor Quantum Dots
By U. Woggon 1996. 80 figs. VIII, 256 pages

Printing: Mercedesdruck, Berlin
Binding: Buchbinderei Lüderitz & Bauer, Berlin